Engineering Drawing

Engineering Drawing

Naseer Khan

RANDOM PUBLICATIONS

NEW DELHI - 110 002 (INDIA)

Engineering Drawing

ISBN 978-93-51112-92-1

Published in 2014 in India by

RANDOM PUBLICATIONS

4376-A/4B, Gali Murari Lal, Ansari Road

New Delhi-110 002

Phone: +9111-43580356, 23289044

E-mail: randomexports@gmail.com; sales@randompublications.com; info@randompublications.com

Reprinted 2026

Type Setting by: Friends Media, Delhi-110089

Preface

For centuries, until the post-World War II era, all engineering drawing was done manually by using pencil and pen on paper or other substrate. Since the advent of computer-aided design (CAD), engineering drawing has been done more and more in the electronic medium with each passing decade. Today most engineering drawing is done with CAD, but pencil and paper have not disappeared. Some of the tools of manual drafting include pencils, pens and their ink, straightedges, T-squares, French curves, triangles, rulers, protractors, dividers, compasses, scales, erasers, and tacks or push pins. And of course the tools also include drawing boards (drafting boards) or tables. The English idiom "to go back to the drawing board", which is a figurative phrase meaning to rethink something altogether, was inspired by the literal act of discovering design errors during production and returning to a drawing board to revise the engineering drawing. Drafting machines are devices that aid manual drafting by combining drawing boards, straightedges, pantographs, and other tools into one integrated drawing environment. CAD provides their virtual equivalents. Producing drawings usually involves creating an original that is then reproduced, generating multiple copies to be distributed to the shop floor, vendors, company archives, and so on. The classic reproduction methods involved blue and white appearances, which is why engineering drawings were long called, and even today are still often called, "blueprints" or "bluelines", even though those terms are anachronistic from a literal perspective, since most copies of engineering drawings today are made by more modern methods that yield black or multicolour lines on white paper. The more generic term "print" is now in common usage in the U.S. to mean any paper copy of an engineering drawing. In the case of CAD drawings, the original is the CAD file, and the printouts of that file are the "prints".

Engineering drawing and artistic drawing are both types of drawing, and either may be called simply "drawing" when the context is implicit. Engineering drawing shares some traits with artistic drawing in that both create pictures. But whereas the purpose of artistic drawing is to convey emotion or artistic sensitivity in some way, the purpose of engineering drawing is to convey information. One of the corollaries that follows from this fact is that, whereas anyone can appreciate artistic drawing, engineering drawing requires some training to understand; but there is also a high degree of objective commonality in the interpretation. In fact, engineering drawing has evolved into a language that is more precise and unambiguous than natural languages; in this sense it is closer to a programming language in its communication ability. Engineering drawing uses an extensive set of conventions to convey information very precisely, with very little ambiguity. An engineering drawing is a legal document, because it communicates all the needed information about "what is wanted" to the people who will expend resources turning the idea into a reality. It is thus a part of a contract; the purchase order and the drawing together, as well as any ancillary documents, constitute the contract. Thus, if the resulting product is wrong, the worker or manufacturer are protected from liability as long as they have faithfully executed the instructions conveyed by the drawing. If those instructions were wrong, it is the fault of the engineer. Because manufacturing and construction are typically very expensive processes, the question of liability for errors has great legal implications as each party tries to blame the other and assign the wasted cost to the other's responsibility. This is the biggest reason why the conventions of engineering drawing have evolved over the decades toward a very precise, unambiguous state.

It is expected that the book will provide a fund of rich experiences to the students and teachers.

I thank all members of my team who have helped in the preparation of the book. My special thanks go to "Random Publications" who have published the book.

—Naseer Khan

Contents

1

Introduction

Engineering Drawing

Engineering drawing, also known as an Engineering Drafting and a type of technical drawing, is used to fully and clearly define requirements for engineered items. Engineering drawing (the activity) produces engineering drawings (the documents). More than just the drawing of pictures, it is also a language—a graphical language that communicates ideas and information from one mind to another. Most especially, it communicates all needed information from the engineer who *designed* a part to the workers who will *make* it.

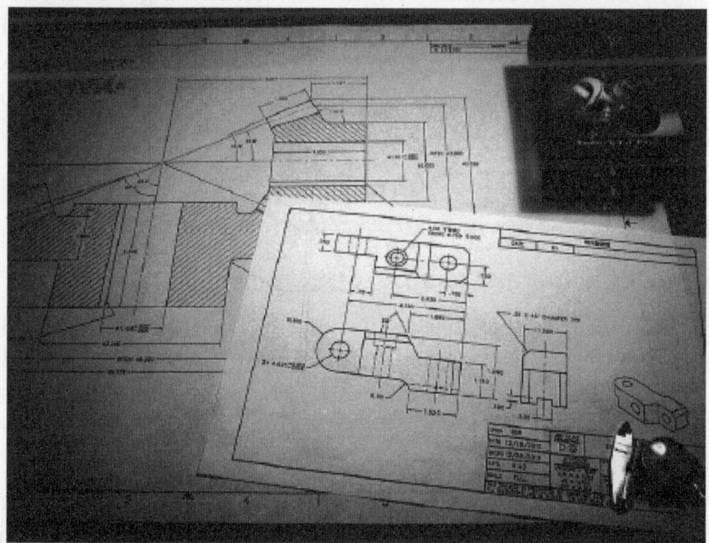

***Figure:** Technical drawings*

Overview

Relationship to Artistic Drawing

Engineering drawing (ABC) and artistic drawing are both types of drawing, and either may be called simply "drawing" when the context is implicit. Engineering drawing shares some traits with artistic drawing in that both create pictures. But whereas the purpose of artistic drawing is to convey emotion or artistic sensitivity in some way (subjective impressions), the purpose of engineering drawing is to convey information (objective facts).

One of the corollaries that follows from this fact is that, whereas anyone can appreciate artistic drawing (even if each viewer has his own unique appreciation), engineering drawing requires some training to understand (like any language); but there is also a high degree of objective commonality in the interpretation (also like other languages). In fact, engineering drawing has evolved into a language that is more precise and unambiguous than natural languages; in this sense it is closer to a programming language in its communication ability. Engineering drawing uses an extensive set of conventions to convey information very precisely, with very little ambiguity. Along with these they are both drawing.

Relationship to other Technical Drawing Types

The process of producing engineering drawings, and the skill of producing them, is often referred to as technical drawing or drafting (also spelled draughting), although technical drawings are also required for disciplines that would not ordinarily be thought of as parts of engineering (such as architecture, landscaping, cabinet making, and garment-making).

Engineering drawing is a universal language.

Cascading of Conventions by Speciality

The various fields share many common conventions of drawing, while also having some field-specific conventions. For example, even within metalworking, there are some process-specific conventions to be learned—casting, machining, fabricating, and assembly all have some special drawing conventions, and within fabrication there is further division, including welding, riveting, pipefitting, and erecting. Each of these trades has some details that only specialists will have memorized.

Legal Instruments

An engineering drawing is a legal document (that is, a legal instrument), because it communicates all the needed information about "what is wanted" to the people who will expend resources turning the idea into a reality. It is thus a part of a contract; the purchase order and the drawing together, as well as any ancillary documents (engineering change orders [ECOs], called-out specs), constitute the contract. Thus, if the resulting product is wrong, the worker or manufacturer are protected from liability as long as they have faithfully executed the instructions conveyed by the drawing. If those instructions were wrong, it is the fault of the engineer. Because manufacturing and construction are typically very expensive processes (involving large amounts of capital and payroll), the question of liability for errors has great legal implications as each party tries to blame the other and assign the wasted cost to the other's responsibility. This is the biggest reason why the conventions of engineering drawing have evolved over the decades towards a very precise, unambiguous state.

Standardization and Disambiguation

In service of the goal of unambiguous communication, engineering drawings are often made professionally and expected to follow certain national and international standards, such as ASME Y14.5 and Y14.5M (most recently revised in 2009) or a group of ISO standards with similar content. Standardization also aids with internationalization, because people from different countries who speak different languages can share the common language of engineering drawing, and can thus communicate with each other quite well, at least as concerns the geometry of an object.

Media

For centuries, until the post-World War II era, all engineering drawing was done manually by using pencil and pen on paper or other substrate (e.g., vellum, mylar). Since the advent of computer-aided design (CAD), engineering drawing has been done more and more in the electronic medium with each passing decade. Today most engineering drawing is done with CAD, but pencil and paper have not disappeared.

Some of the tools of manual drafting include pencils, pens and their ink, straightedges, T-squares, French curves, triangles, rulers, protractors, dividers, compasses, scales, erasers, and tacks or push pins. (Slide rules used to number among the supplies, too, but nowadays even manual drafting, when it occurs, benefits from a pocket calculator or its onscreen equivalent.) And of course the tools also include drawing boards (drafting boards) or tables. The English idiom "to go back to the drawing board", which is a figurative phrase meaning to rethink something altogether, was inspired by the literal act of discovering design errors during production and returning to a drawing board to revise the engineering drawing. Drafting machines are devices that aid manual drafting by combining drawing boards, straightedges, pantographs, and other tools into one integrated drawing environment. CAD provides their virtual equivalents.

Producing drawings usually involves creating an original that is then reproduced, generating multiple copies to be distributed to the shop floor, vendors, company archives, and so on. The classic reproduction methods involved blue and white appearances (whether white-on-blue or blue-on-white), which is why engineering drawings were long called, and even today are still often called, "blueprints" or "bluelines", even though those terms are anachronistic from a literal perspective, since most copies of engineering drawings today are made by more modern methods (often inkjet or laser printing) that yield black or multicolour lines on white paper.

The more generic term "print" is now in common usage in the U.S. to mean any paper copy of an engineering drawing. In the case of CAD drawings, the original is the CAD file, and the printouts of that file are the "prints".

Relationship to Model-based Definition (MBD/DPD)

For centuries, an engineering drawing was the sole method of transferring information from design into manufacture. In recent decades another method has arisen, called model-based definition (MBD) or digital product definition (DPD). In MBD, the information captured by the CAD software app is fed automatically into a CAM app (computer-aided manufacturing), and is translated via postprocessor into other languages such as G-code, which is executed by a CNC machine tool (computer numerical control). Thus today it is often the case that the information travels from the mind of the designer into the manufactured component without having ever been codified by an engineering drawing. In MBD, the dataset, not a drawing, is the legal instrument. The term "technical data package" (TDP) is now used to refer to the *complete package of information* (in one medium or another) that communicates information from design to production (such as 3D-model datasets, engineering drawings, engineering change orders (ECOs), spec revisions and addenda, and so on). However, even in the MBD era, where theoretically production could happen without any drawings or humans at all, it is still the case that drawings and humans are involved.

It still takes CAD/CAM programmers, CNC setup workers, and CNC operators to do manufacturing, as well as other people such as quality assurance staff (inspectors) and logistics staff (for materials handling, shipping-and-receiving, and front office functions).

These workers often use drawings in the course of their work that have been produced by rendering and plotting (printing) from the MBD dataset.

When proper procedures are being followed, a clear chain of precedence is always documented, such that when a person looks at a drawing, s/he is told by a note thereon that this drawing is not the governing instrument (because the MBD dataset is). In these cases, the drawing is still a useful document, although legally it is classified as “for reference only”, meaning that if any controversies or discrepancies arise, it is the MBD dataset, not the drawing, that governs.

Systems of Dimensioning and Tolerancing

Almost all engineering drawings (except perhaps reference-only views or initial sketches) communicate not only geometry (shape and location) but also dimensions and tolerances for those characteristics. Several systems of dimensioning and tolerancing have evolved. The simplest dimensioning system just specifies distances between points (such as an object’s length or width, or hole centre locations). Since the advent of well-developed interchangeable manufacture, these distances have been accompanied by tolerances of the plus-or-minus or min-and-max-limit types.

Coordinate dimensioning involves defining all points, lines, planes, and profiles in terms of Cartesian coordinates, with a common origin. Coordinate dimensioning was the sole best option until the post-World War II era saw the development of geometric dimensioning and tolerancing (GD&T), which departs from the limitations of coordinate dimensioning (e.g., rectangular-only tolerance zones, tolerance stacking) to allow the most logical tolerancing of both geometry and dimensions (that is, both form [shapes/locations] and sizes).

Engineering Drawings: Common Features

Drawings convey the following critical information:

- *Geometry* – the shape of the object; represented as views; how the object will look when it is viewed from various angles, such as front, top, side, etc.
- *Dimensions* – the size of the object is captured in accepted units.
- tolerances – the allowable variations from the nominal size for each dimension.
- *Material* – represents what the item is made of.
- *Finish* – specifies the surface quality of the item, functional or cosmetic. For example, a mass-marketed product usually requires a much higher surface quality than, say, a component that goes inside industrial machinery.

Line Styles and Types

A variety of line styles graphically represent physical objects. Types of *lines* include the following:

- *visible* – are continuous lines used to depict edges directly visible from a particular angle.
- *hidden* – are short-dashed lines that may be used to represent edges that are not directly visible.
- *centre* – are alternately long- and short-dashed lines that may be used to represent the axes of circular features.
- *cutting plane* – are thin, medium-dashed lines, or thick alternately long- and double short-dashed that may be used to define sections for section views.
- *section* – are thin lines in a pattern (pattern determined by the material being "cut" or "sectioned") used to indicate surfaces in section views resulting from "cutting." Section lines are commonly referred to as "cross-hatching."
- *phantom* - (not shown) are alternately long- and double short-dashed thin lines used to represent a feature or component that is not part of the specified part or assembly. E.g. billet ends that may be used for testing, or the machined product that is the focus of a tooling drawing.

Lines can also be classified by a letter classification in which each line is given a letter.

- Type A lines show the outline of the feature of an object. They are the thickest lines on a drawing and done with a pencil softer than HB.
- Type B lines are dimension lines and are used for dimensioning, projecting, extending, or leaders. A harder pencil should be used, such as a 2H.
- Type C lines are used for breaks when the whole object is not shown. They are freehand drawn and only for short breaks. 2H pencil
- Type D lines are similar to Type C, except they are zigzagged and only for longer breaks. 2H pencil
- Type E lines indicate hidden outlines of internal features of an object. They are dotted lines. 2H pencil
- Type F lines are Type F*[typo]* lines, except they are used for drawings in electrotechnology. 2H pencil
- Type G lines are used for centre lines. They are dotted lines, but a long line of 10–20 mm, then a gap, then a small line of 2 mm. 2H pencil
- Type H lines are the same as Type G, except that every second long line is thicker. They indicate the cutting plane of an object. 2H pencil
- Type K lines indicate the alternate positions of an object and the line taken by that object. They are drawn with a long line of 10–20 mm, then a small gap, then a small line of 2 mm, then a gap, then another small line. 2H pencil.

Multiple Views and Projections

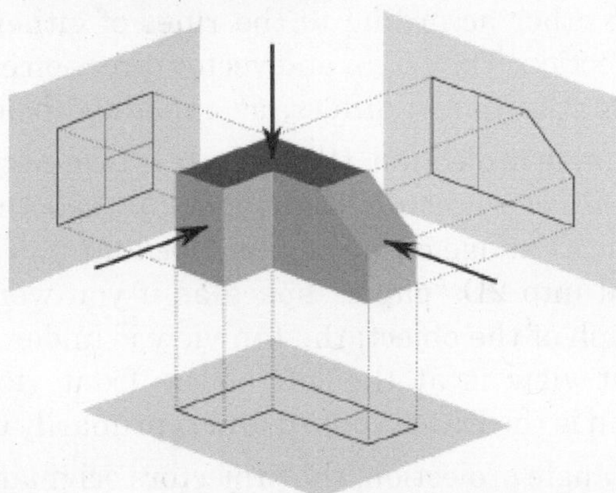

Figure: *Image of a part represented in First Angle Projection*

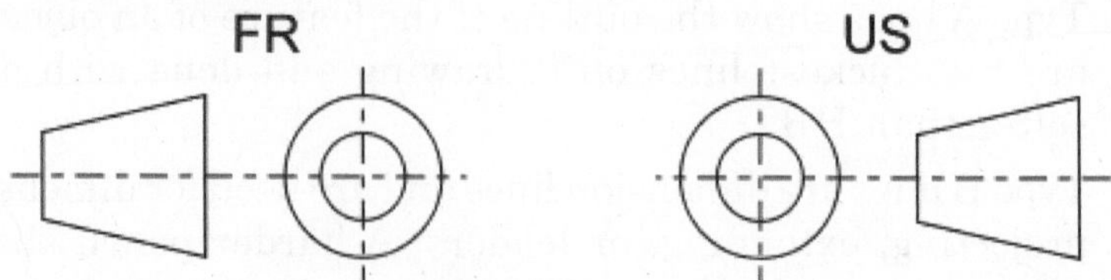

***Figure:** Symbols used to define whether a projection is either Third Angle (right) or First Angle (left).*

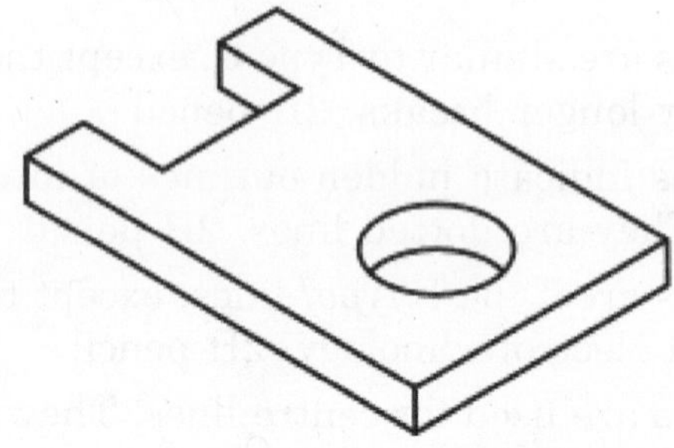

***Figure:** Isometric view of the object shown in the engineering drawing below.*

In most cases, a single view is not sufficient to show all necessary features, and several views are used. Types of *views* include the following:

Orthographic Projection

The orthographic projection shows the object as it looks from the front, right, left, top, bottom, or back, and are typically positioned relative to each other according to the rules of either first-angle or third-angle projection. The origin and vector direction of the projectors (also called projection lines) differs, as explained below.

- In first-angle projection, the projectors originate as if radiated *from a viewer's eyeballs* and shoot *through* the 3D object to project a 2D image onto the plane *behind* it. The 3D object is projected into 2D "paper" space as if you were looking at a radiograph of the object: the top view is under the front view, the right view is at the left of the front view. First-angle projection is the ISO standard and is primarily used in Europe.
- In third-angle projection, the projectors originate as if radiated *from the 3D object itself* and shoot *away from* the 3D object

to project a 2D image onto the plane *in front of* it. The views of the 3D object are like the panels of a box that envelopes the object, and the panels pivot as they open up flat into the plane of the drawing. Thus the left view is placed on the left and the top view on the top; and the features closest to the front of the 3D object will appear closest to the front view in the drawing. Third-angle projection is primarily used in the United States and Canada, where it is the default projection system according to British Standard BS 8888 and ASME standard ASME Y14.3M.

Until the late 19th century, first-angle projection was the norm in North America as well as Europe; but circa the 1890s, the meme of third-angle projection spread throughout the North American engineering and manufacturing communities to the point of becoming a widely followed convention, and it was an ASA standard by the 1950s. Circa World War I, British practice was frequently mixing the use of both projection methods.

As shown above, the determination of what surface constitutes the front, back, top, and bottom varies depending on the projection method used.

Not all views are necessarily used. Generally only as many views are used as are necessary to convey all needed information clearly and economically. The front, top, and right-side views are commonly considered the core group of views included by default, but any combination of views may be used depending on the needs of the particular design. In addition to the 6 principal views (front, back, top, bottom, right side, left side), any auxiliary views or sections may be included as serve the purposes of part definition and its communication. View lines or section lines (lines with arrows marked "A-A", "B-B", etc.) define the direction and location of viewing or sectioning. Sometimes a note tells the reader in which zone(s) of the drawing to find the view or section.

Auxiliary Projection

An auxiliary view is an orthographic view that is projected into any plane other than one of the six principal views. These views are typically used when an object contains some sort of inclined plane. Using the auxiliary view allows for that inclined plane (and any other significant features) to be projected in their true size and shape. The true size and shape of any feature in an engineering drawing can only be known when the Line of Sight (LOS) is perpendicular to the plane being referenced. It is shown like a 3 dimensional object.

Isometric Projection

The isometric projection show the object from angles in which the scales along each axis of the object are equal. Isometric projection corresponds to rotation of the object by ± 45° about the vertical axis, followed by rotation of approximately ± 35.264° [= arcsin(tan(30°))] about the horizontal axis starting from an orthographic projection view. "Isometric" comes from the Greek for "same measure". One of the things that makes isometric drawings so attractive is the ease with which 60 degree angles can be constructed with only a compass and straightedge.

Isometric projection is a type of axonometric projection. The other two types of axonometric projection are:

- Dimetric projection
- Trimetric projection

Oblique Projection

An oblique projection is a simple type of graphical projection used for producing pictorial, two-dimensional images of three-dimensional objects:

- it projects an image by intersecting parallel rays (projectors)
- from the three-dimensional source object with the drawing surface (projection plan).

In both oblique projection and orthographic projection, parallel lines of the source object produce parallel lines in the proj.

Perspective

Perspective is an approximate representation on a flat surface, of an image as it is perceived by the eye. The two most characteristic features of perspective are that objects are drawn:

- Smaller as their distance from the observer increases
- Foreshortened: the size of an object's dimensions along the line of sight are relatively shorter than dimensions across the line of sight.

Section Views

Projected views (either Auxiliary or Orthographic) which show a cross section of the source object along the specified cut plane. These views are commonly used to show internal features with more clarity than may be available using regular projections or hidden lines. In assembly drawings, hardware components (e.g. nuts, screws, washers) are typically not sectioned.

Scale

Plans are usually "scale drawings", meaning that the plans are drawn at specific ratio relative to the actual size of the place or object. Various scales may be used for different drawings in a set. For example, a floor plan may be drawn at 1:50 (1:48 or 1/4"=1'-0") whereas a detailed view may be drawn at 1:25 (1:24 or 1/2"=1'-0"). Site plans are often drawn at 1:200 or 1:100.

Scale is a nuanced subject in the use of engineering drawings. On one hand, it is a general principle of engineering drawings that they are projected using standardized, mathematically certain projection methods and rules. Thus, great effort is put into having an engineering drawing accurately depict size, shape, form, aspect ratios between features, and so on. And yet, on the other hand, there is another general principle of engineering drawing that nearly diametrically opposes all this effort and intent—that is, the principle that *users are not to scale the drawing to infer a dimension not labelled.* This stern admonition is often repeated on drawings, via a boilerplate note in the title block telling the user, "DO NOT SCALE DRAWING."

The explanation for why these two nearly opposite principles can coexist is as follows. The first principle—that drawings will be made so carefully and accurately—serves the prime goal of why engineering drawing even exists, which is successfully communicating part definition and acceptance criteria—including "what the part should look like if you've made it correctly." The service of this goal is what creates a drawing that one even *could* scale and get an accurate dimension thereby. And thus the great temptation to do so, when a dimension is wanted but was not labelled. The second principle—that even though scaling the drawing *will* usually work, one should nevertheless *never* do it—serves several goals, such as enforcing total clarity regarding who has authority to discern design intent, and preventing erroneous scaling of a drawing that was never drawn to scale to begin with (which is typically labelled "drawing not to scale" or "scale: NTS"). When a user is forbidden from scaling the drawing, s/he must turn instead to the engineer (for the answers that the scaling would seek), and s/he will never erroneously scale something that is inherently unable to be accurately scaled. But in some ways, the advent of the CAD and MBD era challenges these assumptions that were formed many decades ago.

When part definition is defined mathematically via a solid model, the assertion that one cannot interrogate the model—the direct analog of "scaling the drawing"—becomes ridiculous; because when part definition is defined this way, it is not *possible* for a drawing or model to be "not to scale".

A 2D pencil drawing can be inaccurately foreshortened and skewed (and thus not to scale), yet still be a completely valid part definition as long as the labelled dimensions are the only dimensions used, and no scaling of the drawing by the user occurs.

This is because what the drawing and labels convey is in reality a *symbol* of what is wanted, rather than a true *replica* of it. (For example, a sketch of a hole that is clearly not round still accurately defines the part as having a true round hole, as long as the label says "10mm DIA", because the "DIA" implicitly but objectively tells the user that the skewed drawn circle is a symbol *representing* a perfect circle.) But if a mathematical model—essentially, a vector graphic—is declared to be the official definition of the part, then any amount of "scaling the drawing" can make sense; there may still be an error in the model, in the sense that what was *intended* is not *depicted* (modelled); but there can be no error of the "not to scale" type—because the mathematical vectors and curves are replicas, not symbols, of the part features.

Even in dealing with 2D drawings, the manufacturing world has changed since the days when people paid attention to the scale ratio claimed on the print, or counted on its accuracy. In the past, prints were plotted on a plotter to exact scale ratios, and the user could know that a line on the drawing 15mm long corresponded to a 30mm part dimension because the drawing said "1:2" in the "scale" box of the title block. Today, in the era of ubiquitous desktop printing, where original drawings or scaled prints are often scanned on a scanner and saved as a PDF file, which is then printed at any percent magnification that the user deems handy (such as "fit to paper size"), users have pretty much given up caring what scale ratio is claimed in the "scale" box of the title block. Which, under the rule of "do not scale drawing", never really did that much for them anyway.

Showing Dimensions

The required sizes of features are conveyed through use of *dimensions*. Distances may be indicated with either of two standardized forms of dimension: linear and ordinate.

- With *linear* dimensions, two parallel lines, called "extension lines," spaced at the distance between two features, are shown at each of the features. A line perpendicular to the extension lines, called a "dimension line," with arrows at its endpoints, is shown between, and terminating at, the extension lines. The distance is indicated numerically at the midpoint of the dimension line, either adjacent to it, or in a gap provided for it.
- With *ordinate* dimensions, one horizontal and one vertical extension line establish an origin for the entire view. The origin is identified with zeroes placed at the ends of these extension lines. Distances along the x- and y-axes to other features are specified using other extension lines, with the distances indicated numerically at their ends.

Sizes of circular features are indicated using either diametral or radial dimensions. Radial dimensions use an "R" followed by the value for the radius; Diametral dimensions use a circle with forward-leaning diagonal line through it, called the *diametre symbol*, followed by the value for the diametre. A radially-aligned line with arrowhead pointing to the circular feature, called a *leader*, is used in conjunction with both diametral and radial dimensions. All types of dimensions are typically composed of two parts: the *nominal* value, which is the "ideal" size of the feature, and the *tolerance*, which specifies the amount that the value may vary above and below the nominal.

- Geometric dimensioning and tolerancing is a method of specifying the functional geometry of an object.

The metric drawing sizes correspond to international paper sizes. These developed further refinements in the second half of the twentieth century, when photocopying became cheap. Engineering drawings could be readily doubled (or halved) in size and put on the next larger (or, respectively, smaller) size of paper with no waste of space. And the metric technical pens were chosen in sizes so that one could add detail or drafting changes with a pen width changing by approximately a factor of the square root of 2. A full set of pens would have the following nib sizes: 0.13, 0.18, 0.25, 0.35, 0.5, 0.7, 1.0, 1.5, and 2.0 mm.

However, the International Organization for Standardization (ISO) called for four pen widths and set a colour code for each: 0.25 (white), 0.35 (yellow), 0.5 (brown), 0.7 (blue); these nibs produced lines that related to various text character heights and the ISO paper sizes.

All ISO paper sizes have the same aspect ratio, one to the square root of 2, meaning that a document designed for any given size can be enlarged or reduced to any other size and will fit perfectly. Given this ease of changing sizes, it is of course common to copy or print a given document on different sizes of paper, especially within a series, e.g. a drawing on A3 may be enlarged to A2 or reduced to A4.

The U.S. customary "A-size" corresponds to "letter" size, and "B-size" corresponds to "ledger" or "tabloid" size. There were also once British paper sizes, which went by names rather than alphanumeric designations.

American Society of Mechanical Engineers (ASME) Y14.2, Y14.3, and Y14.5 are commonly referenced standards in the U.S.

Technical Lettering

Technical lettering is the process of forming letters, numerals, and other characters in technical drawing. It is used to describe, or provide detailed specifications for, an object. With the goals of legibility and uniformity, styles are standardized and lettering ability has little relationship to normal writing ability. Engineering drawings use a Gothic sans-serif script, formed by a series of short strokes. Lower case letters are rare in most drawings of machines. ISO Lettering templates, designed for use with technical pens and pencils, and to suit ISO paper sizes, produce lettering characters to an international standard. The stroke thickness is related to the character height (for example, 2.5mm high characters would have a stroke thickness - pen nib size - of 0.25mm, 3.5 would use a 0.35mm pen and so forth). The ISO character set (font) has a seriffed one, a barred seven, an open four, six, and nine, and a round topped three, that improves legibility when, for example, an A0 drawing has been reduced to A1 or even A3 (and perhaps enlarged back or reproduced/faxed/ microfilmed &c). When CAD drawings became more popular, especially using US American software, such as AutoCAD, the nearest font to this ISO standard font was Romantic Simplex (Roman) - a proprietary shx font) with a manually adjusted width factor (over ride) to make it look as near to the ISO lettering for the drawing board. However, with the closed four, and arced six and nine, romans.shx typeface could be difficult to read in reductions. In more recent revisions of software packages, the TrueType font ISOCPEUR reliably reproduces the original drawing board lettering stencil style, however, many drawings have switched to the ubiquitous Arial.ttf.

Conventional Parts (Areas) of an Engineering Drawing

Responsible dep.	Technical reference	Creator	Approval person		
Wikimedia Commons		Document type P&I Diagram	Document status First issue		
		Title Feed unit with tank	Identification number		
			Rev. 00	Date of issue 2013-01-20	Sheet 1/1

Title block according to EN ISO 7200

Title Block

The title block (T/B, TB) is an area of the drawing that conveys header-type information about the drawing, such as:

- Drawing title (hence the name "title block")
- Drawing number
- Part number(s)
- Name of the design activity (corporation, government agency, etc.)
- Identifying code of the design activity (such as a CAGE code)
- Address of the design activity (such as city, state/province, country)
- Measurement units of the drawing (for example, inches, millimetres)
- Default tolerances for dimension callouts where no tolerance is specified
- Boilerplate callouts of general specs
- Intellectual property rights warning

Traditional locations for the title block are the bottom right (most commonly) or the top right or centre.

Revisions Block

The revisions block (rev block) is a tabulated list of the revisions (versions) of the drawing, documenting the revision control.

Traditional locations for the revisions block are the top right (most commonly) or adjoining the title block in some way.

Effectivity Block

The effectivity block provides a (usually tabular) list of the effectivity of the part design, that is, which higher assemblies it is used in, and thus which models of machine the part is used in. This is sometimes called an Applications list.

Notes List

The notes list provides notes to the user of the drawing, conveying any information that the callouts within the field of the drawing did not. It may include general notes, flagnotes, or a mixture of both. Traditional locations for the notes list are anywhere along the edges of the field of the drawing.

General Notes

General notes (G/N, GN) apply generally to the contents of the drawing, as opposed to applying only to certain part numbers or certain surfaces or features.

Flagnotes

Flagnotes or flag notes (FL, F/N) are notes that apply only where a flagged callout points, such as to particular surfaces, features, or part numbers. Typically the callout includes a flag icon. Some companies call such notes "delta notes", and the note number is enclosed inside a triangular symbol (similar to capital letter delta, Δ). "FL5" (flagnote 5) and "D5" (delta note 5) are typical ways to abbreviate in ASCII-only contexts.

Field of the Drawing

The field of the drawing (F/D, FD) is the main body or main area of the drawing, excluding the title block, rev block, and so on.

List of Materials, Bill of Materials, Parts List

The list of materials (L/M, LM, LoM), bill of materials (B/M, BM, BoM), or parts list (P/L, PL) is a (usually tabular) list of the materials used to make a part, and/or the parts used to make an assembly. It may contain instructions for heat treatment, finishing, and other processes, for each part number. Sometimes such LoMs or PLs are separate documents from the drawing itself. Traditional locations for the LoM/BoM are above the title block, or in a separate document.

Parametre Trabulations

Some drawings call out dimensions with parametre names (that is, variables, such a "A", "B", "C"), then tabulate rows of parametre values for each part number. Traditional locations for parametre tables, when such tables are used, are floating near the edges of the field of the drawing, either near the title block or elsewhere along the edges of the field.

Views and Sections

Each view or section is a separate set of projections, occupying a contiguous portion of the field of the drawing. Usually views and sections are called out with cross-references to specific zones of the field.

Zones

Often a drawing is divided into zones, with labels along the margins, such as A,B,C,D up the sides and 1,2,3,4,5,6 along the top and bottom. Names of zones are thus, for example, A5, D2, or B1.

Abbreviations and Symbols

As in many technical fields, a wide array of abbreviations and symbols have been developed in engineering drawing during the 20th and 21st centuries. For example, cold rolled steel is often abbreviated as CRS, and diametre is often abbreviated as DIA, D, or #.

Example of an Engineering Drawing

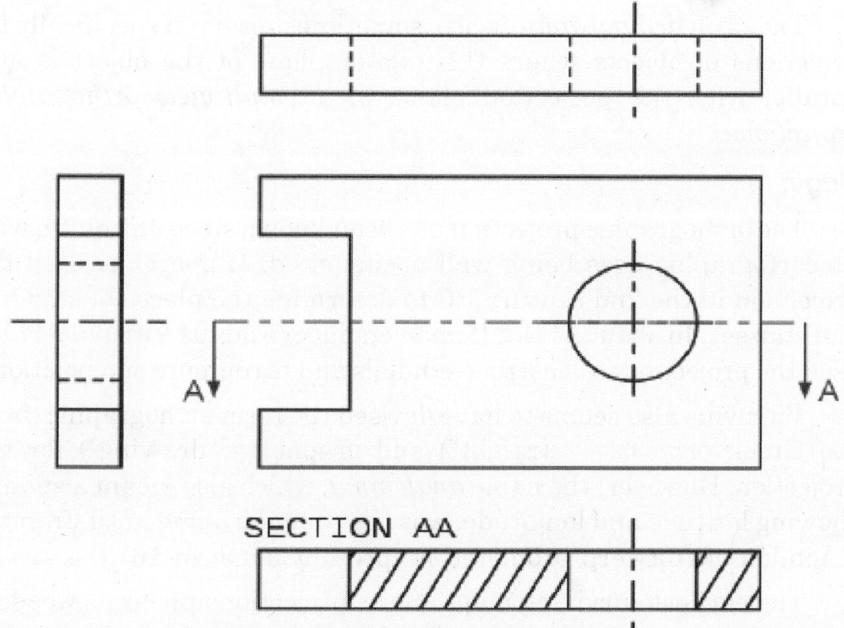

Figure: *Example mechanical drawing*

Here is an example of an engineering drawing (an isometric view of the same object is shown above). The different line types are coloured for clarity.

- Black = object line and hatching
- Red = hidden line
- Blue = centre line of piece or opening
- Magenta = phantom line or cutting plane lines

Sectional views are indicated by the direction of arrows, as in the example above.

Orthographic Projection

Orthographic projection (or orthogonal projection) is a means of representing a three-dimensional object in two dimensions. It is a form of parallel projection, where all the projection lines are orthogonal to the projection plane, resulting in every plane of the scene appearing in affine transformation on the viewing surface.

It is further divided into *multiview orthographic projections* and *axonometric projections*. A lens providing an orthographic projection is known as an (object-space) telecentric lens.

The term *orthographic* is also sometimes reserved specifically for depictions of objects where the axis or plane of the object is also parallel with the projection plane, as in *multiview orthographic projections*.

Origin

The orthographic projection has been known since antiquity, with its cartographic uses being well documented. Hipparchus used the projection in the 2nd century BC to determine the places of star-rise and star-set. In about 14 BC, Roman engineer Marcus Vitruvius Pollio used the projection to construct sundials and to compute sun positions.

Vitruvius also seems to have devised the term orthographic (from the Greek *orthos* (= "straight") and graphç (= "drawing") for the projection. However, the name *analemma*, which also meant a sundial showing latitude and longitude, was the common name until François d'Aguilon of Antwerp promoted its present name in 1613.

The earliest surviving maps on the projection appear as woodcut drawings of terrestrial globes of 1509 (anonymous), 1533 and 1551 (Johannes Schöner), and 1524 and 1551 (Apian).

Multiview Orthographic Projections

With multiview orthographic projections, up to six pictures of an object are produced, with each projection plane parallel to one of the coordinate axes of the object. The views are positioned relative to each other according to either of two schemes: *first-angle* or *third-angle* projection. In each, the appearances of views may be thought of as being *projected* onto planes that form a 6-sided box around the object. Although six different sides can be drawn, *usually* three views of a drawing give enough information to make a 3D object. These views are known as front view, top view and end view.

Axonometric Projection

Axonometric projection is a type of parallel projection used to create a pictorial drawing of an object, where the object is rotated along one or more of its axes relative to the plane of projection.

There are three main types of axonometric projection: *isometric, dimetric*, and *trimetric projection.*

"Axonometric" means "to measure along axes". Axonometric projection shows an image of an object as viewed from a skew direction in order to reveal more than one side in the same picture. Whereas the term *orthographic* is sometimes reserved specifically for depictions of objects where the axis or plane of the object *is* parallel with the projection plane, in axonometric projection the plane or axis of the object is always drawn *not* parallel to the projection plane.

With axonometric projections the scale of distant features is the same as for near features, so such pictures will look distorted, as it is not how our eyes or photography work. This distortion is especially evident if the object to view is mostly composed of rectangular features. Despite this limitation, axonometric projection can be useful for purposes of illustration.

The concept of an isometric projection had existed in a rough empirical form for centuries, well before Professor William Farish (1759–1837) of Cambridge University was the first to provide detailed rules for isometric drawing.

Farish published his ideas in the 1822 paper "On Isometrical Perspective", in which he recognized the "need for accurate technical working drawings free of optical distortion. This would lead him to formulate isometry. Isometry means "equal measures" because the same scale is used for height, width, and depth".

From the middle of the 19th century, according to Jan Krikke (2006) isometry became an "invaluable tool for engineers, and soon thereafter axonometry and isometry were incorporated in the curriculum of architectural training courses in Europe and the U.S. The popular acceptance of axonometry came in the 1920s, when modernist architects from the Bauhaus and De Stijl embraced it". De Stijl architects like Theo van Doesburg used axonometry for their architectural designs, which caused a sensation when exhibited in Paris in 1923".

Since the 1920s axonometry, or parallel perspective, has provided an important graphic technique for artists, architects, and engineers. Like linear perspective, axonometry helps depict 3D space on the 2D picture plane. It usually comes as a standard feature of CAD systems and other visual computing tools.

According to Jan Krikke (2000) however, "axonometry originated in China. Its function in Chinese art was similar to linear perspective in European art. Axonometry, and the pictorial grammar that goes with it, has taken on a new significance with the advent of visual computing".

Three Types of Axonometric Projections

The three types of axonometric projections are isometric projection, *dimetric projection*, and *trimetric projection*, depending on the exact angle at which the view deviates from the orthogonal. Typically in axonometric drawing, one axis of space is shown as the vertical.

- In isometric projection, the most commonly used form of axonometric projection in engineering drawing, the direction of viewing is such that the three axes of space appear equally foreshortened, and there is a common angle of 60° between them. As the distortion caused by foreshortening is uniform the proportionality of all sides and lengths are preserved, and the axes share a common scale. This enables measurements to be read or taken directly from the drawing. Another advantage is that 60° angles are more easily constructed using only a compass and straightedge.
- In *dimetric projection*, the direction of viewing is such that two of the three axes of space appear equally foreshortened, of which the attendant scale and angles of presentation are determined according to the angle of viewing; the scale of the third direction (vertical) is determined separately.

- In *trimetric projection*, the direction of viewing is such that all of the three axes of space appear unequally foreshortened. The scale along each of the three axes and the angles among them are determined separately as dictated by the angle of viewing. Trimetric perspective is seldom used, and is found in only a few video games (*Fallout*, *SimCity 4*).

Approximations are common in dimetric and trimetric drawings.

Limitations of Axonometric Projection

s with all types of parallel projection, objects drawn with axonometric projection do not appear larger or smaller as they extend closer to or away from the viewer.

While advantageous for architectural drawings, where measurements must be taken directly from the image, the result is a perceived distortion, since unlike perspective projection, this is not how photography normally works. It also can easily result in situations where depth and altitude are difficult to gauge, as is shown in the illustration to the right.

In this isometric drawing, the blue sphere is two units higher than the red one. However, this difference in elevation is not apparent if one covers the right half of the picture, as the boxes (which serve as clues suggesting height) are then obscured.

This visual ambiguity has been exploited in op art, including "impossible object" drawings. M. C. Escher's *Waterfall* (1961) is a well-known example, in which a channel of water seems to travel unaided along a downward path, only to then paradoxically fall once again as it returns to its source. The water thus appears to disobey the law of conservation of energy.

2

Multiview Orthographic Projection

In technical drawing and computer graphics, a multiview orthographic projection is an illustration technique in which up to six pictures of an object are produced, with each projection plane parallel to one of the coordinate axes of the object. The views are positioned relative to each other according to either of two schemes: *first-angle* or *third-angle* projection. In each, the appearances of views may be thought of as being *projected* onto planes that form a 6-sided box around the object.

Quadrants in Descriptive Geometry

Modern orthographic projection is derived from Gaspard Monge's descriptive geometry. Monge defined a reference system of two viewing planes, horizontal *H* ("ground") and vertical *V* ("backdrop"). These two planes intersect to partition 3D space into 4 quadrants, which he labelled:

- *I*: above *H*, in front of *V*
- *II*: above *H*, behind *V*
- *III*: below *H*, behind *V*
- *IV*: below *H*, in front of *V*

These quadrant labels are the same as used in 2D planar geometry, as seen from infinitely far to the "left", taking *H* and *V* to be the *X*-axis and *Y*-axis, respectively. The 3D object of interest is then placed into either quadrant *I* or *III* (equivalently, the position of the intersection line between the two planes is shifted), obtaining *first-* and *third-angle* projections, respectively.

Quadrants *II* and *IV* are also mathematically valid, but their use would result in one view "true" and the other view "flipped" by 180° through its vertical centerline, which is too confusing for technical drawings. (In cases where such a view is useful, e.g. a ceiling viewed from above, a reflected view is used, which is a mirror image of the true orthographic view.)

Monge's original formulation uses two planes only, and obtains the top and front views only. The addition of a third plane to show a side view (either left or right) is a modern extension. The terminology of *quadrant* is a mild anachronism, as a modern orthographic projection with three views corresponds more precisely to an octant of 3D space.

First-angle Projection

In first-angle projection, the object is conceptually located in quadrant I, i.e. it floats above and before the viewing planes, the planes are opaque, and each view is pushed through the object onto the plane furthest from it. (Mnemonic: an "actor on a stage".) Extending to the 6-sided box, each view of the object is projected in the direction (sense) of sight of the object, onto the (opaque) interior walls of the box; that is, each view of the object is drawn on the opposite side of the box. A two-dimensional representation of the object is then created by "unfolding" the box, to view all of the interior walls. This produces two plans and four elevations. A simpler way to visualize this is to place the object on top of an upside-down bowl. Sliding the object down the right edge of the bowl reveals the right side view. Also, if you get two types of cars this will give us a better view of the car.

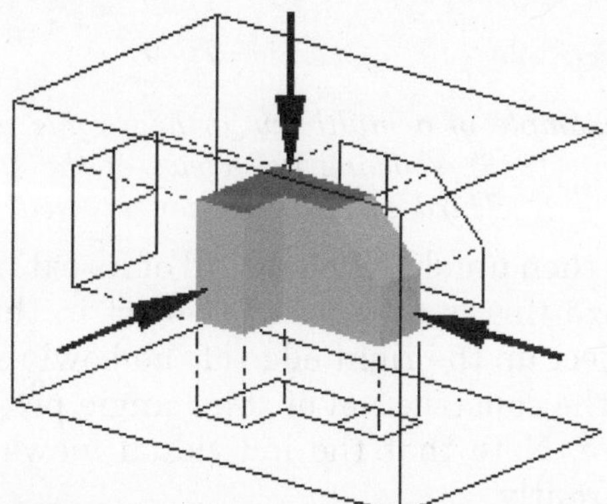

Figure: *Image of object in box, with views of object projected in the direction of sight onto walls using first-angle projection.*

Third-angle Projection

In third-angle projection, the object is conceptually located in quadrant III, i.e. it is positioned below and behind the viewing planes, the planes are transparent, and each view is pulled onto the plane closest to it. (Mnemonic: a "shark in a tank", esp. that is sunken into the floor.) Using the 6-sided viewing box, each view of the object is projected opposite to the direction (sense) of sight, onto the (transparent) exterior walls of the box; that is, each view of the object is drawn on the same side of the box.

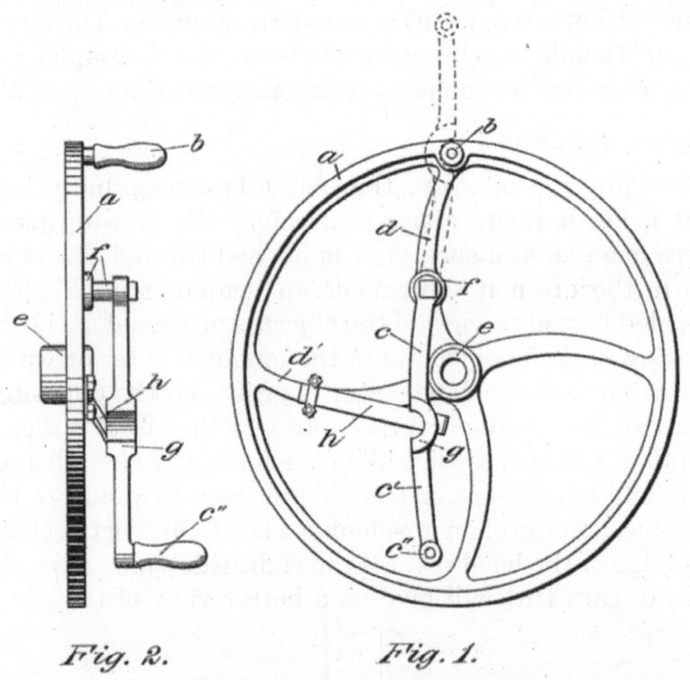

Figure: *An example of a multiview orthographic drawing from a US Patent (1913), showing two views of the same object. Third angle projection is used.*

The box is then unfolded to view all of its exterior walls. A simpler way to visualize this is to place the object in the bottom of a bowl. Sliding the object up the right edge of the bowl reveals the right side view. Here is the construction of third angle projections of the same object as above. Note that the individual views are the same, just arranged differently.

Additional Information

First-angle projection is as if the object were sitting on the paper and, from the "face" (front) view, it is rolled to the right to show the left side or rolled up to show its bottom. It is standard throughout Europe (excluding the UK) and Asia. First-angle projection used to be common in the UK, and may still be seen on historical design drawings, but has now fallen into disuse in favour of third-angle projection. Third-angle is as if the object were a box to be unfolded. If we unfold the box so that the front view is in the centre of the two arms, then the top view is above it, the bottom view is below it, the left view is to the left, and the right view is to the right. It is standard in the United Kingdom (BS 8888:2006 specifies it as the default projection system), USA (ASME Y14.3-2003 specifies it as the default projection system), Canada, and Australia.

Both first-angle and third-angle projections result in the same 6 views; the difference between them is the arrangement of these views around the box.

A great deal of confusion has ensued in drafting rooms and engineering departments when drawings are transferred from one convention to another. On engineering drawings, the projection angle is denoted by an international symbol consisting of a truncated cone, respectively for first-angle (FR) and third-angle (US):

The 3D interpretation of the symbol can be deduced by envisioning a solid truncated cone, standing upright with its large end on the floor and the small end upward. The top view is therefore two concentric circles ("donut"). In particular, the fact that the inner circle is drawn with a solid line instead of dashed disambiguates this view as the top view, not the bottom view.

- In *first-angle projection*, the "top" view is pushed down to the floor, and the "front" view is pushed back to the rear wall; the intersection line between these two planes is therefore closest to the large end of the cone, hence the first-angle symbol shows the cone with its large end open towards the donut.
- In *third-angle projection*, the "top" view is pulled up to the ceiling, and the "front" view is pulled forward to the front wall; the intersection line between the two planes is thus closest to the small end of the cone, hence the third-angle symbol shows the cone with its large end away from the donut.

Multiviews Without Rotation

Orthographic multiview projection is derived from the principles of descriptive geometry and may produce an image of a specified, imaginary object as viewed from any direction of space. Orthographic projection is distinguished by parallel projectors emanating from all points of the imaged object and which intersect a plane of projection at right angles. Above, a technique is described that obtains varying views by projecting images after the object is rotated to a desired position. Descriptive geometry customarily relies on obtaining various views by imagining an object to be stationary, and changing the direction of projection (viewing) in order to obtain the desired view.

Using the rotation technique above, note that no orthographic view is available looking perpendicularly at any of the inclined surfaces. Suppose a technician desired such a view to, say, look through a hole to be drilled perpendicularly to the surface. Such a view might be desired for calculating clearances or for dimensioning purposes. To obtain this view without multiple rotations requires the principles of Descriptive Geometry. The steps below describe the use of these principles in third angle projection.

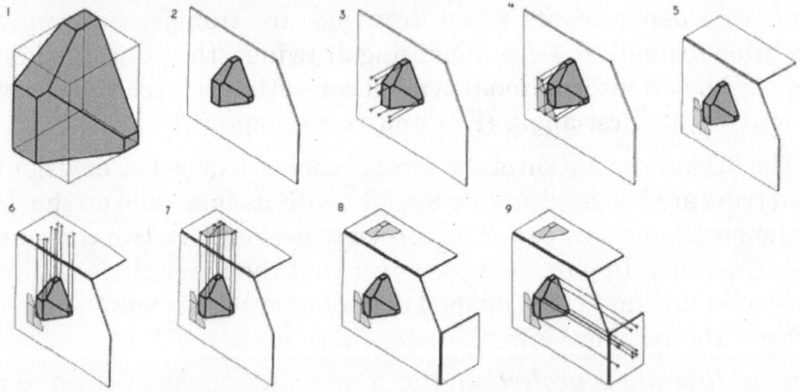

***Figure:** Figures one through nine.*

- *Fig.1*: Pictorial of imaginary object that the technician wishes to image.
- *Fig.2*: The object is imagined behind a vertical plane of projection. The angled corner of the plane of projection is addressed later.
- *Fig.3*: Projectors emanate parallel from all points of the object, perpendicular to the plane of projection.
- *Fig.4*: An image is created thereby.
- *Fig.5*: A second, horizontal plane of projection is added, perpendicular to the first.
- *Fig.6*: Projectors emanate parallel from all points of the object perpendicular to the second plane of projection.
- *Fig.7*: An image is created thereby.

- *Fig.8*: A third plane of projection is added, perpendicular to the previous two.
- *Fig.9*: Projectors emanate parallel from all points of the object perpendicular to the third plane of projection.

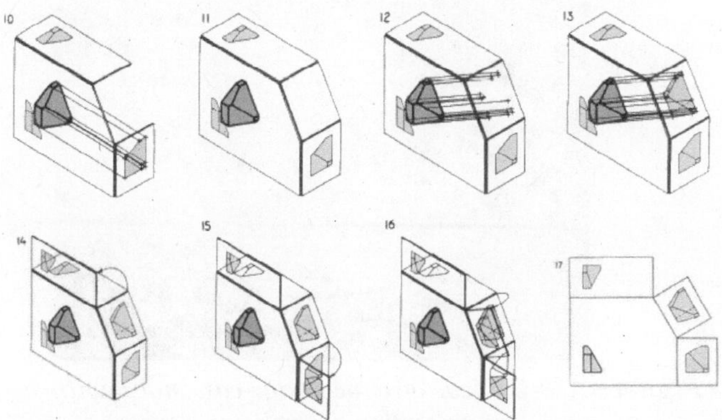

Figure: *Figures ten through seventeen.*

- *Fig.10*: An image is created thereby.
- *Fig.11*: A fourth plane of projection is added parallel to the chosen inclined surface, and per force, perpendicular to the first (Frontal) plane of projection.
- *Fig.12*: Projectors emanate parallel from all points of the object perpendicularly from the inclined surface, and per force, perpendicular to the fourth (Auxiliary) plane of projection.
- *Fig.13*: An image is created thereby.
- *Fig.14-16*: The various planes of projection are unfolded to be planar with the Frontal plane of projection.
- *Fig.17*: The final appearance of an orthographic multiview projection and which includes an "Auxiliary view" showing the true shape of an inclined surface.

Views

Section

A section, or *cross-section*, is a view of a 3-dimensional object from the position of a plane through the object. A cross section is a common method of depicting the internal arrangement of a 3-dimensional object in two dimensions. It is often used in technical drawing and is traditionally crosshatched. The style of crosshatching indicates the type of material the section passes through.

With computed axial tomography, computers construct cross-sections from x-ray data.

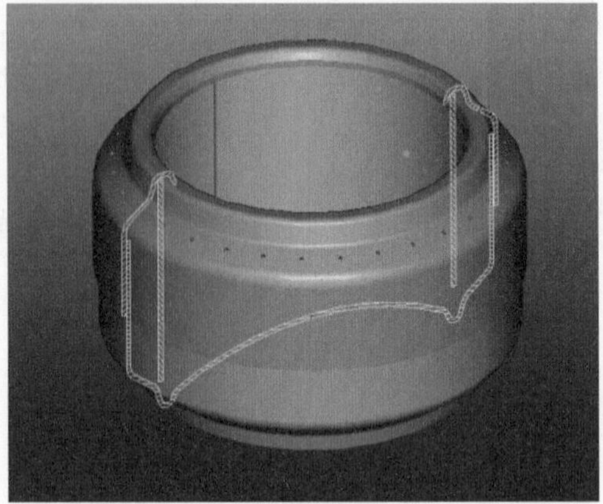

***Figure:** A 3-D view of a beverage-can stove with a cross-section in yellow.*

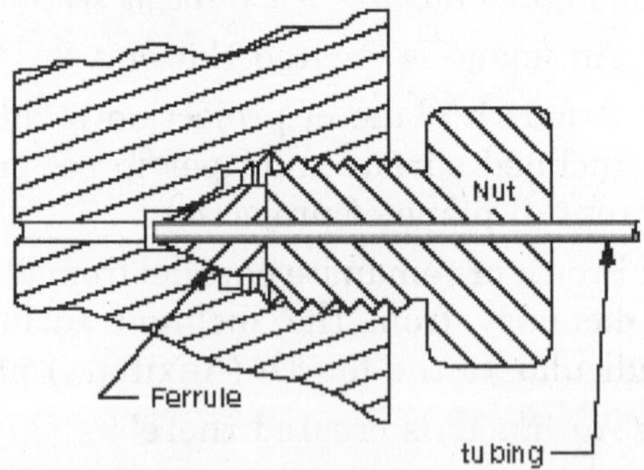

***Figure:** A 2-D cross-sectional view of a compression seal.*

***Figure:** Half-section of a Porsche 996*

Elevation

Figure: *Principal façade of the Panthéon, Paris, by Jacques-Germain Soufflot.*

An elevation is a view of a 3-dimensional object from the position of a horizontal plane beside an object. In other words, an elevation is a side-view as viewed from the front, back, left or right (and referred to as a *front elevation, [left/ right] side elevation*, and a *rear elevation*). It is the corollary to the concept of a "view" (which is always overhead and is therefore referred to as an *overhead view*).

An elevation is a common method of depicting the external configuration and detailing of a 3-dimensional object in two dimensions. Building façades are shown as elevations in architectural drawings and technical drawings.

Elevations are the most common orthographic projection for conveying the appearance of a building from the exterior. Perspectives are also commonly used for this purpose. A building elevation is typically labelled in relation to the compass direction it faces; the direction from which a person views it. E.g. the North Elevation of a building is the side that most closely faces true north on the compass. Interior elevations are used to show detailing such as millwork and trim configurations.

In the building industry elevations are a non-perspective view of the structure. These are drawn to scale so that measurements can be taken for any aspect necessary. Drawing sets include front, rear and both side elevations. The elevations specify the composition of the different facades of the building, including ridge heights, the positioning of the final fall of the land, exterior finishes, roof pitches and other architectural details.

Developed Elevation

A developed elevation is a variant of a regular elevation view in which several adjacent non-parallel sides may be shown together, as if they have been unfolded. For example, the north and west views may be shown side-by-side, sharing an edge, even though this does not represent a proper orthographic projection.

Plan

A plan is a view of a 3-dimensional object from the position of a horizontal plane through, above, or below the object. In such views, the portion of the object in front of the plane is omitted to reveal what lies beyond. In the case of a floor plan, the roof and upper portion of the walls may be omitted. Elevations, top (roof) plans, and bottom plans are orthographic projections, but they are not sections as their viewing plane is outside of the object. A plan is a common method of depicting the internal arrangement of a 3-dimensional object in two dimensions. It is often used in technical drawing and is traditionally cross-hatched. The style of crosshatching indicates the type of material the section passes through.

Auxiliary View

An auxiliary view is a view taken from an angle that is *not* one of the primary views. An auxiliary view is a view at an angle used to give deeper insight into the actual shape of the object. An auxiliary view is used to show a slanted surface in true size and shape. This is accomplished by providing a view that is perpendicular to the slanted surface. These allow the true shape/dimension of features at any angle relative to the main views to be shown .

Graphical Projection

Graphical projection is a protocol by which an image of a three-dimensional object is projected onto a planar surface without the aid of mathematical calculation, used in technical drawing.

Overview

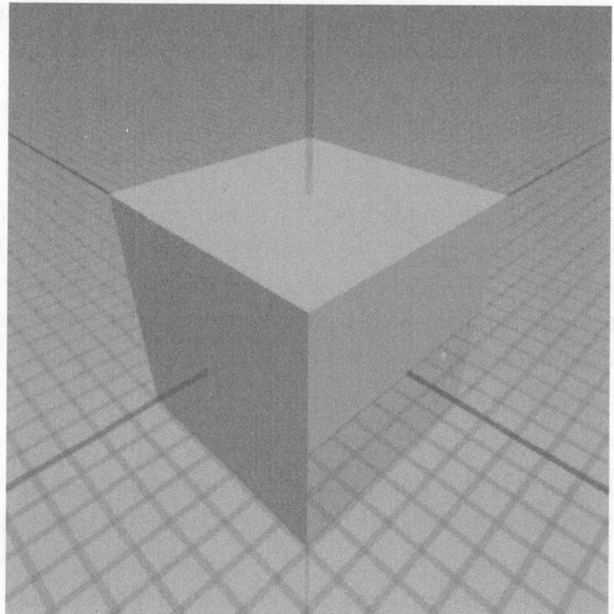

Figure: *Increasing the focal length and distance of the camera to infinity in a perspective projection results in a parallel projection.*

The projection is achieved by the use of imaginary "projectors". The projected, mental image becomes the technician's vision of the desired, finished picture. By following the protocol the technician may produce the envisioned picture on a planar surface such as drawing paper. The protocols provide a uniform imaging procedure among people trained in technical graphics (mechanical drawing, computer aided design, etc.).

There are two graphical projection categories each with its own protocol:

- parallel projection
- perspective projection

Types of Projection

Parallel Projection

In parallel projection, the lines of sight from the object to the projection plane are parallel to each other. Within parallel projection there is an ancillary category known as "pictorials". Pictorials show an image of an object as viewed from a skew direction in order to reveal all three directions (axes) of space in one picture. Because pictorial projections innately contain this distortion, in the rote, drawing instrument for pictorials, some liberties may be taken for economy of effort and best effect. It is a simultaneous process of viewing the image give pictures

Orthographic Projection

The Orthographic projection is derived from the principles of descriptive geometry and is a two-dimensional representation of a three-dimensional object. It is a parallel projection (the lines of projection are parallel both in reality and in the projection plane). It is the projection type of choice for working drawings.

Pictorials

Within parallel projection there is a subcategory known as *Pictorials*. Pictorials show an image of an object as viewed from a skew direction in order to reveal all three directions (axes) of space in one picture.

Parallel projection pictorial instrument drawings are often used to approximate graphical perspective projections, but there is attendant distortion in the approximation.

Because pictorial projections inherently have this distortion, in the instrument drawing of pictorials, great liberties may then be taken for economy of effort and best effect. Parallel projection pictorials rely on the technique of axonometric projection ("to measure along axes").

Axonometric Projection

Axonometric projection is a type of parallel projection used to create a pictorial drawing of an object, where the object is rotated along one or more of its axes relative to the plane of projection.

There are three main types of axonometric projection: *isometric*, *dimetric*, and *trimetric projection.*

Isometric Projection

In isometric pictorials, the direction of viewing is such that the three axes of space appear equally foreshortened, and there is a common angle of 60° between them. As the distortion caused by foreshortening is uniform the proportionality of all sides and lengths are preserved, and the axes share a common scale. This enables measurements to be read or taken directly from the drawing.

Dimetric Projection

In dimetric pictorials, the direction of viewing is such that two of the three axes of space appear equally foreshortened, of which the attendant scale and angles of presentation are determined according to the angle of viewing; the scale of the third direction (vertical) is determined separately. Approximations are common in dimetric drawings.

Trimetric Projection

In trimetric pictorials, the direction of viewing is such that all of the three axes of space appear unequally foreshortened. The scale along each of the three axes and the angles among them are determined separately as dictated by the angle of viewing. Approximations in Trimetric drawings are common.

Oblique Projection

In oblique projections the parallel projection rays are not perpendicular to the viewing plane as with orthographic projection, but strike the projection plane at an angle other than ninety degrees. In both orthographic and oblique projection, parallel lines in space appear parallel on the projected image. Because of its simplicity, oblique projection is used exclusively for pictorial purposes rather than for formal, working drawings.

In an oblique pictorial *drawing*, the displayed angles among the axes as well as the foreshortening factors (scale) are arbitrary. The distortion created thereby is usually attenuated by aligning one plane of the imaged object to be parallel with the plane of projection thereby creating a true shape, full-size image of the chosen plane. Special types of oblique projections are:

Cavalier Projection

In cavalier projection (sometimes cavalier perspective or high view point) a point of the object is represented by three coordinates, x, y and z. On the drawing, it is represented by only two coordinates, x'' and y''.

On the flat drawing, two axes, x and z on the figure, are perpendicular and the length on these axes are drawn with a 1:1 scale; it is thus similar to the dimetric projections, although it is not an orthographic projection, as the third axis, here y, is drawn in diagonal, making an arbitrary angle with the x'' axis, usually 30 or 45°. The length of the third axis is not scaled.

Cabinet Projection

The term cabinet projection (sometimes cabinet perspective) stems from its use in illustrations by the furniture industry. Like cavalier perspective, one face of the projected object is parallel to the viewing plane, and the third axis is projected as going off in an angle (typical 30° or 45°). Unlike cavalier projection, where the third axis keeps its length, with cabinet projection the length of the receding lines is cut in half.

Perspective Projection

Perspective projection is a linear projection where three dimensional objects are projected on a *picture plane*. This has the effect that distant objects appear smaller than nearer objects.

It also means that lines which are parallel in nature (that is, meet at the point at infinity) appear to intersect in the projected image, for example if railways are pictured with perspective projection, they appear to converge towards a single point, called vanishing point. Photographic lenses and the human eye work in the same way, therefore perspective projection looks most realistic. Perspective projection is usually categorized into one-point, two-point and three-point perspective, depending on the orientation of the projection plane towards the axes of the depicted object.

Graphical projection methods rely on the duality between lines and points, whereby two straight lines determine a point while two points determine a straight line. The orthogonal projection of the eye point onto the picture plane is called the *principal vanishing point* (P.P. in the scheme on the left, from the Italian term *punto principale*, coined during the renaissance).

Two relevant points of a line are:

- its intersection with the picture plane, and
- its vanishing point, found at the intersection between the parallel line from the eye point and the picture plane.

The principal vanishing point is the vanishing point of all horizontal lines perpendicular to the picture plane. The vanishing points of all horizontal lines lie on the horizon line, obviously. If, as is often the case, the picture plane is vertical, all vertical lines are drawn vertically, and have no finite vanishing point on the picture plane. Various graphical methods can be easily envisaged for projecting geometrical scenes. For example, lines traced from the eye point at 45° to the picture plane intersect the latter along a circle whose radius is the distance of the eye point from the plane, thus tracing that circle aids the construction of all the vanishing points of 45° lines; in particular, the intersection of that circle with the horizon line consists of two *distance points*.

They are useful for drawing chessboard floors which, in turn, serve for locating the base of objects on the scene. In the perspective of a geometric solid on the right, after choosing the principal vanishing point —which determines the horizon line— the 45° vanishing point on the left side of the drawing completes the characterization of the (equally distant) point of view.

Two lines are drawn from the orthogonal projection of each vertex, one at 45° and one at 90° to the picture plane. After intersecting the ground line, those lines go towards the distance point (for 45°) or the principal point (for 90°). Their new intersection locates the projection of the map. Natural heights are measured above the ground line and then projected in the same way until they meet the vertical from the map.

Isometric Projection

Isometric projection is a method for visually representing three-dimensional objects in two dimensions in technical and engineering drawings. It is an axonometric projection in which the three coordinate axes appear equally foreshortened and the angles between any two of them are 120 degrees.

Overview

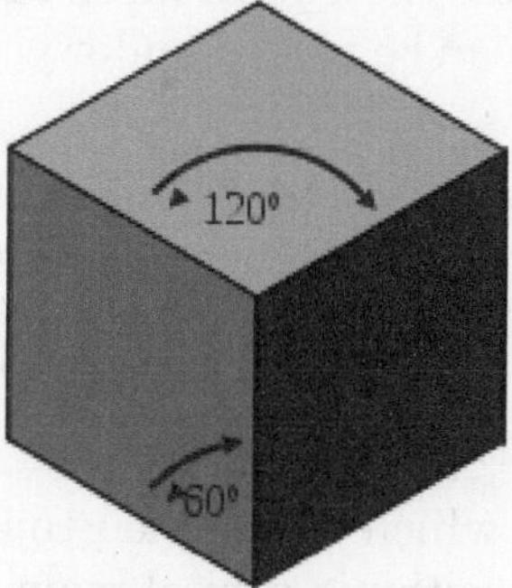

Figure: *Isometric drawing of a cube.*

The term "isometric" comes from the Greek for "equal measure", reflecting that the scale along each axis of the projection is the same (unlike some other forms of graphical projection).

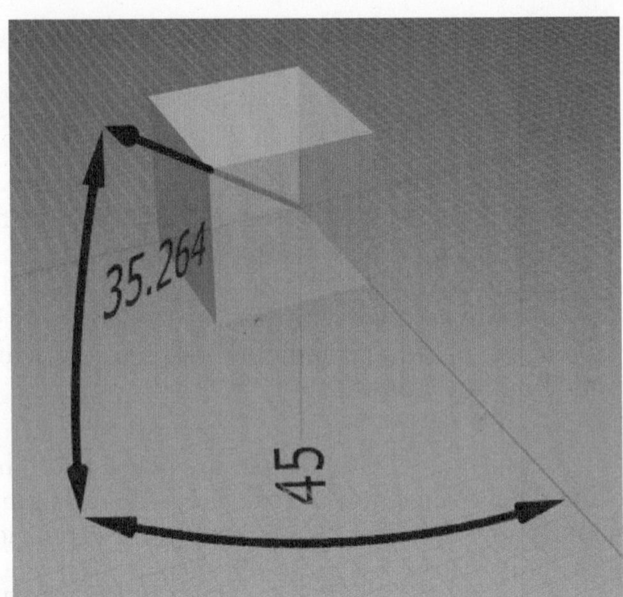

Figure: *Camera rotations needed to achieve this perspective.*

An isometric view of an object can be obtained by choosing the viewing direction in a way that the angles between the projection of the x, y, and z axes are all the same, or 120°. For example when taking a cube, this is done by first looking straight towards one face. Next the cube is rotated ±45° about the vertical axis, followed by a rotation of approximately ±35.264° (precisely arcsin(tan 30°) or arctan(sin 45°)) about the horizontal axis. Note that with the cube the perimetre of the 2D drawing is a perfect regular hexagon: all the black lines are of equal length and all the cube's faces are the same area. There is isometric paper that can be placed under your normal piece of drawing paper as an aid.

In a similar way an *isometric view* can be obtained for example in a 3D scene editor. Starting with the camera aligned parallel to the floor and aligned to the coordinate axes, it is first rotated downwards around the horizontal axis by about 35.264° as above, and then rotated ±45° around the vertical axis.

Another way in which isometric projection can be visualized is by considering a view within a cubical room starting in an upper corner and looking towards the opposite, lower corner. The x-axis extends diagonally down and right, the y-axis extends diagonally down and left, and the z-axis is straight up. Depth is also shown by height on the image. Lines drawn along the axes are at 120° to one another.

The term "isometric" is often mistakenly used to refer to axonometric projections in general. (There are three types of axonometric projections: isometric, dimetric and trimetric.)

Rotation Angles

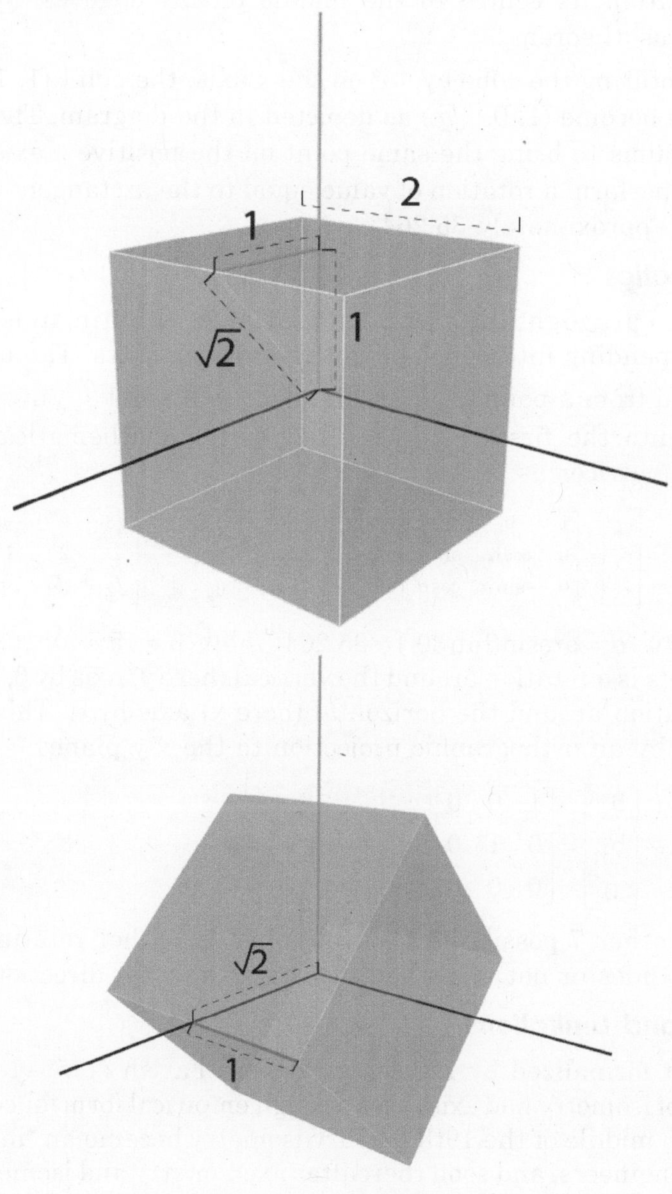

Figure: *Rotate on y for* $arctan(1/\sqrt{2})$

From the two angles needed for an isometric projection, the value of the second may seem counter intuitive and deserves some further explanation. Let's first imagine a cube with sides of length 2, and its centre positioned at the axis origin. We can calculate the length of the line from its centre to the middle of any edge as $\sqrt{2}$ using Pythagoras' theorem .

By rotating the cube by 45° on the x axis, the point (1, 1, 1) will therefore become (1, 0, $\sqrt{2}$) as depicted in the diagram. The second rotation aims to bring the same point on the positive z axis and so needs to perform a rotation of value equal to the arctangent of $1/\sqrt{2}$ which is approximately 35.264°.

Mathematics

There are eight different orientations to obtain an isometric view, depending into which octant the viewer looks. The isometric transform from a point $a_{x,y,z}$ in 3D space to a point $b_{x,y}$ in 2D space looking into the first octant can be written mathematically with rotation matrices as:

$$\begin{bmatrix} c_x \\ c_y \\ c_z \end{bmatrix} = \begin{bmatrix} 1 & 0 & 0 \\ 0 & \cos\alpha & \sin\alpha \\ 0 & -\sin\alpha & \cos\alpha \end{bmatrix} \begin{bmatrix} \cos\beta & 0 & -\sin\beta \\ 0 & 1 & 0 \\ \sin\beta & 0 & \cos\beta \end{bmatrix} \begin{bmatrix} a_x \\ a_y \\ a_z \end{bmatrix} = \frac{1}{\sqrt{6}} \begin{bmatrix} \sqrt{3} & 0 & -\sqrt{3} \\ 1 & 2 & 1 \\ \sqrt{2} & -\sqrt{2} & \sqrt{2} \end{bmatrix} \begin{bmatrix} a_x \\ a_y \\ a_z \end{bmatrix}$$

where $\alpha = \arcsin(\tan 30^\circ) \approx 35.264^\circ$ and $\beta = 45^\circ$. As explained above, this is a rotation around the vertical (here y) axis by β, followed by a rotation around the horizontal (here x) axis by α. This is then followed by an orthographic projection to the x-y plane:

$$\begin{bmatrix} b_x \\ b_y \\ 0 \end{bmatrix} = \begin{bmatrix} 1 & 0 & 0 \\ 0 & 1 & 0 \\ 0 & 0 & 0 \end{bmatrix} \begin{bmatrix} c_x \\ c_y \\ c_z \end{bmatrix}$$

The other 7 possibilities are obtained by either rotating to the opposite sides or not, and then inverting the view direction or not.

History and Limitations

First formalized by Professor William Farish (1759–1837), the concept of isometry had existed in a rough empirical form for centuries. From the middle of the 19th century isometry became an "invaluable tool for engineers, and soon thereafter axonometry and isometry were incorporated in the curriculum of architectural training courses in Europe and the U.S."

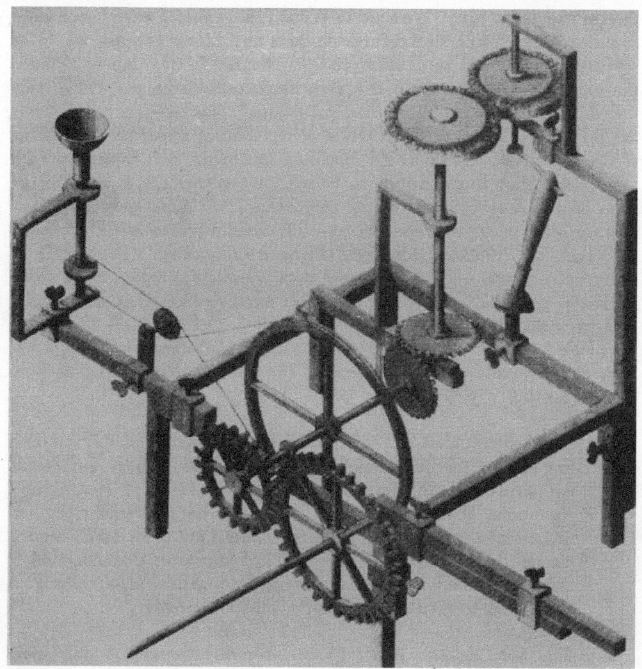

***Figure:** Optimal-grinding engine model (1822), drawn in 30° isometric.*

***Figure:** Example of Chinese art in an illustrated edition of the* Romance of the Three Kingdoms, *China, c. 15th century.*

According to Jan Krikke (2000) however, "axonometry originated in China. Its function in Chinese art was similar to linear perspective in European art. Axonometry, and the pictorial grammar that goes with it, has taken on a new significance with the advent of visual computing".

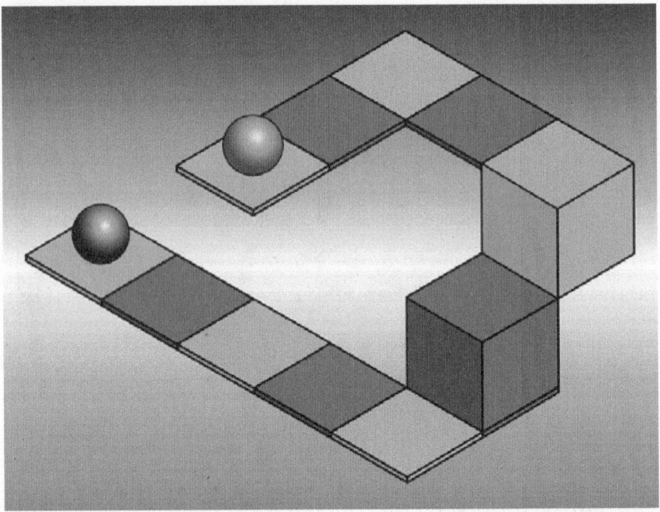

Figure: *An example of the limitations of isometric projection. The height difference between the red and blue balls cannot be determined locally.*

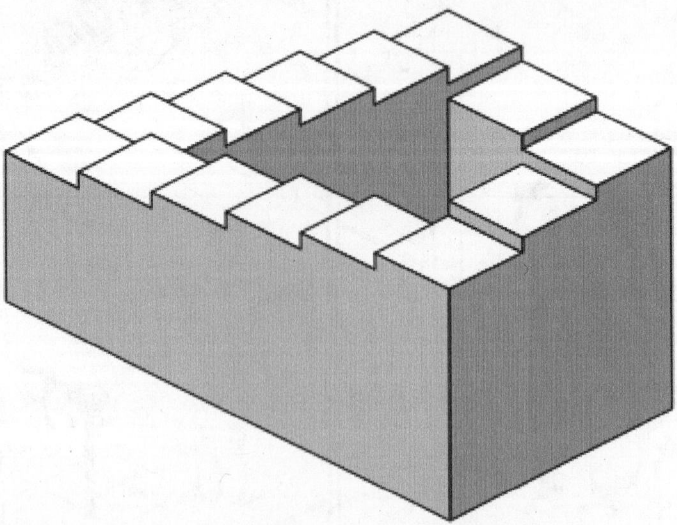

Figure: *The Penrose stairs depicts a staircase which seems to ascend (anticlockwise) or descend (clockwise) yet forms a continuous loop.*

As with all types of parallel projection, objects drawn with isometric projection do not appear larger or smaller as they extend closer to or away from the viewer. While advantageous for architectural drawings where measurements need to be taken directly, the result is a perceived distortion, as unlike perspective projection, it is not how our eyes or photography normally work. It also can easily result in situations where depth and altitude are difficult to gauge, as is shown in the illustration to the right.

Architect's Scale

An architect's scale is a specialized ruler designed to facilitate the drafting and measuring of architectural drawings, such as floor plans and orthographic projections. Because the scale of such drawings are often smaller than life-size, an architect's scale features multiple units of length and proportional length increments.

For accuracy and longevity, the material used should be dimensionally stable and durable. Scales were traditionally made of wood, but today they are usually made of rigid plastic or aluminium. Architect's scales may be flat, with 4 scales, or have a symmetric 3-lobed cross-section, with 6 scales.

United States and Imperial Units

In the United States, and prior to metrification in Britain, Canada and Australia, architect's scales are/were marked as a ratio of x inches-to-the-foot (typically written as x"=1'-0"). For example one inch measured from a drawing with a scale of "one-inch-to-the-foot" is equivalent to one foot in the real world (a scale of 1:12) whereas one inch measured from a drawing with a scale of "two-inches-to-the-foot" is equivalent to six inches in the real world (a scale of 1:6). It is not to be confused with a true unitless ratio (inches to inches)— a 1:5 architectural scale would be a 1:60 unitless scale.

Typical scales used in the United States are:

- Full scale, with inches divided into sixteenths of an inch

The following scales are generally grouped in pairs using the same dual-numbered index line (one scale is read from the right, and the other scale is read from the left):

- three-inches-to-the-foot (3"=1'-0") (ratio equivalent 1:4)/one-and-one-half-inch-to-the-foot (1½"=1'-0") (1:8)
- one-inch-to-the-foot (1"=1'-0") (1:12)/one-half-inch-to-the-foot (1/2"=1'-0") (1:24)

- three-quarters-inch-to-the-foot (3/4"=1'-0") (1:16)/three-eighths-inch-to-the-foot (3/8"=1'-0") (1:32)
- one-quarter-inch-to-the-foot (1/4"=1'-0") (1:48)/one-eighth-inch-to-the-foot (1/8"=1'-0") (1:96)
- three-sixteenths-inch-to-the-foot (3/16"=1'-0") (1:64)/three-thirty-seconds-inch-to-the-foot (3/32"=1'0") (1:128)

Metric Units

Architect's scale rulers used in Britain and other metric countries are marked with ratios without reference to a base unit. Therefore a drawing will indicate both its scale (ratio) and the unit of measurement being used. In Britain, and elsewhere, the standard units used on architectural drawings are the (SI) units millimetres (mm) and metres (m), whereas in France centimetres (cm) and metres are most often used.

In Britain, for flat rulers, the paired scales often found on architect's scales are:

- 1:1/1:100
- 1:5/1:50
- 1:10/1:100
- 1:20/1:200
- 1:1250/1:2500

For triangular rulers, the paired scales are:

- 1:1/1:10
- 1:2/1:20
- 1:5/1:50
- 1:100/1:200
- 1:500/1:1000
- 1:1250/1:2500

Less common scales are:

- 1:25/1:250
- 1:33^{1}D

Engineer's Scale

Any scale is called an engineer's scale and is a tool for measuring distances and transferring measurements at a fixed ratio of length. It is commonly made of plastic,or aluminum and is just over 12 inches

(305 mm) long, but with only 12 inches of markings, leaving the ends unmarked so that the first and last measuring ticks do not wear off. It is used in making engineering drawings, commonly called blueprints, blue lines or plans in a specific scale. For example, "one-tenth size" would appear on a drawing to indicate a part larger than the drawing on the paper itself. It is not to be used to measure machined parts to see if they meet specifications.

In scientific and engineering terminology, a device to measure linear distance and create proportional linear measurements is called a scale. A device for drawing straight lines is a straight edge or ruler. In common usage both are referred to as a ruler.

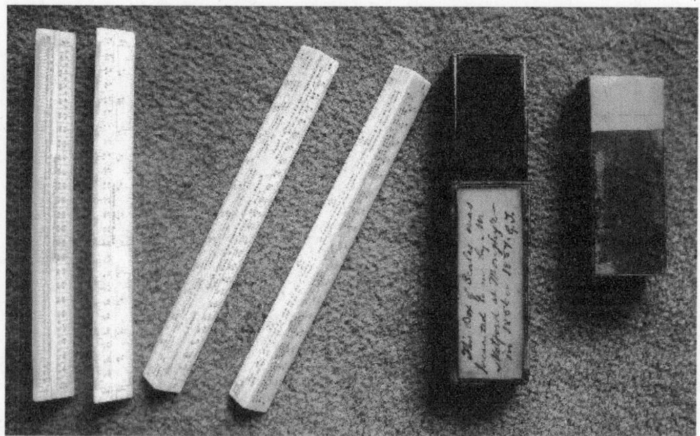

***Figure:** Boxed set of 1850s ivory engineer's scales presented to the railway civil engineer George Turnbull in India. 16 scales are engraved.*

In Canada and the United States, this scale is divided into decimalized fractions of an inch, but has a cross-section like an equilateral triangle, which enables the scale to have six edges indexed for measurement. One edge is divided into tenths of an inch, and the subsequent ones are directly marked for twentieths, thirtieths, fortieths, fiftieths, and finally sixtieths of an inch. Referred to as 1:10, 1:20, 1:30,1:40, 1:50 or 1:60 scale. Typically in civil engineering applications, 1:10 (1"=10')is used exclusively for detail drawings. 1:20 and 1:40 scales are used for working plans. 1:60 is normally used only to show large areas of a project.

The engineer's scale came into existence when machining parts required a greater precision than the usual, binary fractionalization of the inch as in the architect's scale for houses and furniture. They were used, for example, in laying out printed circuit boards with the spacing of leads from integrated circuits as one-tenth of an inch.

On the other hand, they may have come into existence far earlier for use in civil engineering applications when project scope could far exceed the potential of an architect's scale.

Geometric Dimensioning and Tolerancing

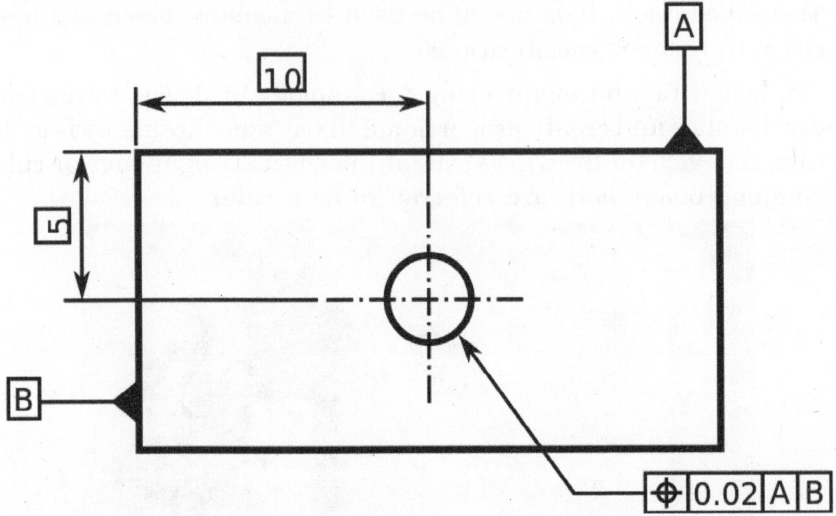

Figure: *Example of geometric dimensioning and tolerancing*

Geometric dimensioning and tolerancing (GD&T) is a system for defining and communicating engineering tolerances. It uses a symbolic language on engineering drawings and computer-generated three-dimensional solid models that explicitly describes nominal geometry and its allowable variation.

It tells the manufacturing staff and machines what degree of accuracy and precision is needed on each controlled feature of the part. GD&T is used to define the nominal (theoretically perfect) geometry of parts and assemblies, to define the allowable variation in form and possible size of individual features, and to define the allowable variation between features.

- Dimensioning specifications define the nominal, as-modelled or as-intended geometry. One example is a basic dimension.
- Tolerancing specifications define the allowable variation for the form and possibly the size of individual features, and the allowable variation in orientation and location between features. Two examples are linear dimensions and feature control frames using a datum reference (both shown above).

There are several standards available worldwide that describe the symbols and define the rules used in GD&T. One such standard is American Society of Mechanical Engineers (ASME) Y14.5-2009. This article is based on that standard, but other standards, such as those from the International Organization for Standardization (ISO), may vary slightly. The Y14.5 standard has the advantage of providing a fairly complete set of standards for GD&T in one document. The ISO standards, in comparison, typically only address a single topic at a time. There are separate standards that provide the details for each of the major symbols and topics below (e.g. position, flatness, profile, etc.).

Dimensioning and Tolerancing Philosophy

According to the ASME Y14.5-2009 standard, the purpose of geometric dimensioning and tolerancing (GD&T) is to describe the engineering intent of parts and assemblies. This is not a completely correct explanation of the purpose of GD&T or dimensioning and tolerancing in general.

The purpose of GD&T is more accurately defined as describing the geometric requirements for part and assembly geometry. Proper application of GD&T will ensure that the allowable part and assembly geometry defined on the drawing leads to parts that have the desired form and fit (within limits) and function as intended.

There are some fundamental rules that need to be applied (these can be found on page 7 of the 2009 edition of the standard):

- All dimensions must have a tolerance. Every feature on every manufactured part is subject to variation, therefore, the limits of allowable variation must be specified. Plus and minus tolerances may be applied directly to dimensions or applied from a general tolerance block or general note. For basic dimensions, geometric tolerances are indirectly applied in a related Feature Control Frame. The only exceptions are for dimensions marked as minimum, maximum, stock or reference.
- Dimensioning and tolerancing shall completely define the nominal geometry and allowable variation. Measurement and scaling of the drawing is not allowed except in certain cases.
- Engineering drawings define the requirements of finished (complete) parts. Every dimension and tolerance required to define the finished part shall be shown on the drawing. If additional dimensions would be helpful, but are not required, they may be marked as reference.

- Dimensions should be applied to features and arranged in such a way as to represent the function of the features. Additionally, dimensions should not be subject to more than one interpretation.
- Descriptions of manufacturing methods should be avoided. The geometry should be described without explicitly defining the method of manufacture.
- If certain sizes are required during manufacturing but are not required in the final geometry (due to shrinkage or other causes) they should be marked as non-mandatory.
- All dimensioning and tolerancing should be arranged for maximum readability and should be applied to visible lines in true profiles.
- When geometry is normally controlled by gage sizes or by code (e.g. stock materials), the dimension(s) shall be included with the gage or code number in parentheses following or below the dimension.
- Angles of 90° are assumed when lines (including centre lines) are shown at right angles, but no angular dimension is explicitly shown. (This also applies to other orthogonal angles of 0°, 180°, 270°, etc.)
- Dimensions and tolerances are valid at 20 °C / 101.3 kPa unless stated otherwise.
- Unless explicitly stated, all dimensions and tolerances are only valid when the item is in a free state.
- Dimensions and tolerances apply to the full length, width, and depth of a feature including form variation.
- Dimensions and tolerances only apply at the level of the drawing where they are specified. It is not mandatory that they apply at other drawing levels, unless the specifications are repeated on the higher level drawing(s).

(Note: The rules above are not the exact rules stated in the ASME Y14.5-2009 standard.)

GD&T Data Exchange

Exchange of geometric dimensioning and tolerancing (GD&T) information between CAD systems is available on different levels of fidelity for different purposes:

- In the early days of CAD exchange only lines, texts and symbols were written into the exchange file. A receiving system could display them on the screen or print them out, but only a human could interpret them.
- *GD&T presentation*: On a next higher level the presentation information is enhanced by grouping them together into *callouts* for a particular purpose, e.g. a *datum feature callout* and a *datum reference frame*.

  And there is also the information which of the curves in the exchange file are leader, projection or dimension curves and which are used to form the shape of a product.
- *GD&T representation*: Unlike GD&T presentation, the GD&T representation does not deal with how the information is presented to the user but only deal with which element of a shape of a product has which GD&T characteristic.

  A system supporting GD&T representation may display the GD&T information in some tree and other dialogs and allow the user to directly select and highlight the corresponding feature on the shape of the product, 2D and 3D.
- Ideally both GD&T presentation and representation are available in the exchange file and are associated with each other.

  Then a receiving system can allow a user to select a GD&T callout and get the corresponding feature highlighted on the shape of the product.
- An enhancement of GD&T representation is defining a formal language for GD&T (similar to a programming language) which also has built-in rules and restrictions for the proper GD&T usage. This is still a research area.
- *GD&T validation*: Based on GD&T representation data (but not on GD&T presentation) and the shape of a product in some useful format (e.g. a boundary representation), it is possible to validate the completeness and consistency of the GD&T information. The software tool FBTol from the Kansas City Plant is probably the first one in this area.
- GD&T representation information can also be used for the software assisted manufacturing planning and cost calculation of parts.

GD&T & ISO TC 10 Technical Product Documentation

- ISO 128 *Technical drawings – Indication of dimensions and tolerances*
- ISO 7083 *Symbols for geometrical tolerancing – Proportions and dimensions*
- ISO 13715 *Technical drawings – Edges of undefined shape – Vocabulary and indications*
- ISO 15786 *Simplified representation and dimensioning of holes*
- ISO 16792:2006 *Technical product documentation—Digital product definition data practices* (Note: ISO 16792:2006 was derived from ASME Y14.41-2003 by permission of ASME)
- ISO 2010
- ISO

ISO/TC 213 Dimensional and Geometrical Product Specifications and Verification

In ISO/TR 14638 *GPS – Masterplan* the distinction between fundamental, global, general and complementary GPS standards is made.

- Fundamental GPS standards
  - ISO 8015 *Concepts, principles and rules*
- Global GPS standards
  - ISO 14660-1 *Geometrical features*
  - ISO/TS 17, orientation and location
  - ISO 1101 *Geometrical tolerancing – Tolerances of form, orientation, location and run-out*
    - Amendment 1 *Representation of specifications in the form of a 3D model*
  - ISO 1119 *Series of conical tapers and taper angles*
  - ISO 2692 *Geometrical tolerancing – Maximum material requirement (MMR), least material requirement (LMR) and reciprocity requirement (RPR)*
  - ISO 3040 *Dimensioning and tolerancing – Cones*
  - ISO 5458 *Geometrical tolerancing – Positional tolerancing*
  - ISO 5459 *Geometrical tolerancing – Datums and datum systems*

    - ISO 10578 *Tolerancing of orientation and location – Projected tolerance zone*
    - ISO 10579 *Dimensioning and tolerancing – Non-rigid parts*
    - ISO 14406 *Extraction*
    - ISO 22432 *Features utilized in specification and verification*
- General GPS standards: Areal and profile surface texture
    - ISO 1302 *Indication of surface texture in technical product documentation*
    - ISO 3274 *Surface texture: Profile method – Nominal characteristics of contact (stylus) instruments*
    - ISO 4287 *Surface texture: Profile method – Terms, definitions and surface texture parametres*
    - ISO 4288 *Surface texture: Profile method – Rules and procedures for the assessment of surface texture*
    - ISO 8785 *Surface imperfections – Terms, definitions and parametres*
    - Form of a surface independent of a datum or datum system. Each of them has a part 1 for the *Vocabulary and parametres* and a part 2 for the *Specification operators*:
        - ISO 12180 *Cylindricity*
        - ISO 12181 *Roundness*
        - ISO 12780 *Straightness*
        - ISO 12781 *Flatness*
    - ISO 25178 *Surface texture: Areal*
- General GPS standards: Extraction and filtration techniques
    - ISO/TS 1661 *Filtration*
    - ISO 11562 *Surface texture: Profile method – Metrological characteristics of phase correct filters*
    - ISO 12085 *Surface texture: Profile method – Motif parametres*
    - ISO 13565 *Profile method; Surfaces having stratified functional properties*

ASME *Standards* American Society of Mechanical Engineers

- ASME Y14.41-2003 Digital Product Definition Data Practices
- ASME Y14.5 - 2009 Dimensioning and Tolerancing

- ASME Y14.5M-1994 Dimensioning and Tolerancing
- ASME Y14.5.1M-1994 Mathematical Definition of Dimensioning and Tolerancing Principles

ASME Y14.5.1M-2

GD&T Standards for Data Exchange and Integration

- ISO 10303 *Industrial automation systems and integration — Product data representation and exchange*
  - o ISO 10303-47 *Integrated generic resource: Shape variation tolerances*
  - o ISO/TS 10303-1130 *Application module: Derived shape element*
  - o ISO/TS 10303-1050 *Application module: Dimension tolerance*
  - o ISO/TS 10303-1051 *Application module: Geometric tolerance*
  - o ISO/TS 10303-1052 *Application module: Default tolerance*
  - o ISO/TS 10303-1666 *Application module: Extended geometric tolerance*
  - o ISO 10303-203 *Application protocol: Configuration controlled 3D design of mechanical parts and assemblies*
  - o ISO 10303-210 *Application protocol: Electronic assembly, interconnection, and packaging design*
  - o ISO 10303-214 *Application protocol: Core data for automotive mechanical design processes*
  - o ISO 10303-224 *Application protocol: Mechanical product definition for process planning using machining features*
  - o ISO 10303-238 *Application protocol: Application interpreted model for computerized numerical controllers* (STEP-NC)

Layout of Drawing Sheets

All engineering drawings have to be drawn strictly on drawing sheets of standard sizes.

The use of standard sizes for drawing sheets saves paper and ensures convenient storage of drawing. Layout of drawing sheets has been discussed of Basic Engineering Drawing. The conventions and specifications of standard formats of sheet sizes and some important aspects of layout are discussed in this chapter, as recommended by Bureau of Indian Standard.

Layout of Drawing Sheet

The layout of a drawing sheet is the standard arrangement to ensure that all the necessary information is included in the drawing sheet to facilitate its quick reading and easy reference. The requirements of a good layout of drawing sheets are:

(a) It should facilitate the reading of drawings.

(b) Essential references are located.

(c) It includes all necessary informations.

(d) It must provide extra margin for filing.

The width of the grid reference system shall be approximately half the width of the border for a given size of sheet. The length of each grid shall not be less than 25 mm and not more than 75 mm. The rectangles of the grid should be denoted by means of capital letters along both the vertical edges and numerals along both the horizontal edges. The numbering and lettering starts at the sheet corner opposite to the title block and is repeated on the opposite sides. The letters and numerals are to be written upright. If the number of the lettered divisions exceeds that of the alphabet, the reference letters with the extra division should be doubled, for example, AA, BB, CC... etc. The layout of a drawing sheet also includes the standard practices of placement of drawing sheets, border sizes to be left for different sizes of drawing sheets, the filing margin, the details to be included in the title block, the grid reference system etc.

Item List or Parts List

An *item list* or a *parts list* is a complete list of all the associated parts constituting an assembly or a sub-assembly. This may be presented on a technical drawing itself or provided as a separate document. When the item list is included on the drawing itself, it is to be placed in the viewing direction of the drawing and in conjunction with the title block. The sequence of entries must be from bottom to top with the headings of the columns at the bottom immediately above title block. When the item list is shown as a separate document, the sequence of entries must be from top to bottom with the headings of the columns at the top. When the item list is shown as a separate document, it must be given the same number as that of the parent drawing. However, to distinguish it from that of the parent drawing, it is recommended that the item list number be preceded by the prefix "Item list". The sheet sizes for preparing the item list must be chosen from the ISO – A series given in table.

Entries in the Item list. The standard details to be entered in the item list are discussed below.

(a) *Item:* The "Item" column shows the relevant item number as shown on the relevant drawing.

(b) *Reference* The "Reference" column is used to identify items which are not completely represented on the parent drawing such as parts represented on other drawings, standard parts, or other readymade parts. According to the case, the number of the other drawings, their relevant standard and the code or any similar information may be entered here.

(c) *Description:* In the "Description" column, the designation of the item must be entered. Abbreviations may be used if they do not impair clarity. If the item concerns a standard part, for example, bolt, nut, stud, washer, etc., its standard designation must be used in accordance with the relevant standard.

(d) *Quantity* In the "Quantity" column, the total number of that particular item necessary for one complete assembly must be entered.

(e) *Material* In the "Material" column, the type and the quality of the material to be used must be entered. If standard material is used, its standard designation must be entered.

The sequence of these entries in the item list is optional. Each of these details has to be arranged in columns. If more details other than the standard details are required, more columns may be added to cover the specific requirement.

Item Reference

Item reference is assigning of numbers in sequential order to each of the component part of the assembly. Identical parts shown in the same assembly must have the same item reference number. The item references must be generally written in Hindu-Arabic numerals only. They may, however, be augmented by capital letters when necessary. The height of the item reference shall be generally twice the height of characters used in dimensioning. For the sake of clarity and legibility of the drawing, item references must be arranged in vertical columns or horizontal rows. The item references of related items may be shown against the same leader with circles touching each other.

The sequence of numbering may be according to the possible order of the assembly, or according to the importance of the component parts, or according to any other logical sequence.

Engineering Drawing Dimensions

An engineering drawing, besides illustrating the shape of an object, must also show the size and any other relevant details essential for the construction of the object. Indicating on a drawing, the sizes of the object and other details essential for its construction and function using lines, numerals, symbols, notes, etc., is called *dimensioning*. The dimensions are given to indicate the sizes of various features of the object and their locations. Indicated dimensions of an object should not be mistaken as those which are required to make the drawing. The dimensions indicated on engineering drawings are those which are essential for production, inspection and functioning of the object.

Method of Dimensioning

The two systems of placing the dimensions on a drawing are *Aligned System* and *Unidirectional System*. Both the systems are recommended in *SP: 46-1988* by Bureau of Indian Standards and have been taught in Basic Engineering Drawing.

Method of Execution:

Dimensioning in engineering drawing is executed as follows:

(a) Dimensioning lines and projection lines (extension lines) are drawn as continuous thin lines.

(b) Extension lines should be extended slightly beyond dimension lines.

(c) The construction lines and the intersecting projection lines should extend slightly beyond their point of intersection.

(d) Projection and dimension lines should not cross other lines, unless this is unavoidable.

(e) Projection lines are drawn in a direction perpendicular to the feature to be dimensioned, or where necessary, they may be drawn obliquely but parallel to each other.

(f) In Aligned System of dimensioning, except where unavoidable, dimension lines should not be placed in the 30° zone.

Leaders

Leaders (or *pointer lines*) are continuous thin lines drawn from notes and figures to show where they apply. Leaders are terminated either by arrow heads or dots. While arrow heads connected to a leader should always terminate at a line, the dots connected to a leader should be within the outline of the object drawn freehand.

Leaders may also terminate in a short horizontal bar at the bottom level of the lettering of the first or last line of the note. Leaders which touch lines should not normally be inclined at an angle less than 30^0, and they should not be parallel to adjacent dimension or projection line, where confusion might arise. The use of common leaders should be avoided even if it means repeating the dimensions or notes.

Arrowhead

Each end of a dimension line should be defined by an arrowhead, whose two branches should be at an angle sufficiently open to mark clearly the extremities of the dimension lines. Alternatively, the length of the arrow head may be taken as about three times the width. The space in the arrowhead is filled in. The size of arrowheads should be proportionate to the thickness of the lines of the drawing and size of drawing. It is recommended that the barbs of the arrowhead form an included angle of 15^0 or more.

Arrowheads should be drawn within the limits of the dimensioned feature. When enough space is not available, they should be placed outside.

Addition of Letters and Symbols

The following letters or symbols are commonly used in dimensioning:

(a) Before a dimension for a diametre, the symbol 'Φ' should be placed.

(b) Before a dimension for a radius, the letter R should be written.

(c) Before a dimensions for a square, the letters SQ should be written.

(d) Before across dimensions for a hexagon, the letters HEX should be written. If it clearly appears from the drawing that the dimension applies to a square section or hexagonal section the letter SQ or HEX may be omitted.

(e) The radius or diametre dimension of a spherical surface should be preceded by the word SPHERE.

Angular Dimensioning

Angular dimensions and their deviations are dimensioned in the same manner as that of dimensioning linear dimensions. However, angular dimensions should be placed. In certain cases, these may be written horizontally if this improves clarity.

Arrangement of Dimensions

(a) *Chain Dimensioning*: Chain dimensioning should only be used where the possible accumulation of tolerances does not endanger the functional requirement of the part.

(b) *Parallel Dimensioning*: Where a number of dimensions have a common datum line/surface, the dimensions are indicated by individual parallel dimension lines.

(c) *Combined Dimensioning* Combined dimensioning results from this simultaneous use of the two methods i.e., chain and parallel dimensioning.

(d) *Progressive Dimensioning*: In parallel dimensioning, since the Dimension lines of each of the dimensions are drawn separately parallel to each other, a large space is required for dimensioning. Where there is space constraint or likelihood of confusion, *progressive dimensioning* may be used for simplicity. In this method, the datum line is indicated by a dot and a zero sign. The dimensions and the zero are placed in line with the projection lines.

(e) *Dimensioning by Co-ordinates:* In some cases, it may be useful, instead of dimensioning to group dimensions separately from the drawing.

(f) *Symmetrical, Equal or Equidistant Dimensioning.* When a dimension is divided into several parts, the sign equal to '=' may be used to indicate those dimensions which are normally equal. Where equidistant or regularly arranged elements appear on a drawing, the method may be used for simplicity.

(g) *Assembled Parts Dimensioning:* When several parts are drawn in assembly, the groups of dimensions related to each part shall be kept separate as much as possible.

(h) Dimensions, the true length of which is not possible to be shown in any view, may be indicated. This method is to be used only when it is not practicable to provide sufficient view and sections to show dimensions in the true length.

Engineering Drawing Sectional View

An object or an assembly of parts will show its inner details when cut away at a proper place. The imaginary cutting of a part or an assembly of parts in engineering drawing is known as sectioning and the views obtained are called the sectional views. The object is sectioned in such a way as to show the parts and inner details as clearly as possible, which otherwise cannot be visualized even with the hidden lines.

Sectional Views

Sectioning is usually performed on the object for the generally accepted orthographic views viz. the front view, side view and top view. Sectioning is done by an imaginary cutting plane, on the object, parallel to that particular view. The cutting plane, in most of the cases passes through the centre of the symmetric object but is shifted if the inner details become more clear at some other point. The cutting-plane line should be named with capital letters of double the height as used for dimension. The arrowheads at the end of the line show the direction of the viewer.

Types of Section There are Mainly Seven Types of Sections

(a) Full section
(b) Half section
(c) Partial or Broken out section
(d) Offset section
(e) Rotated or Revolved section
(f) Removed section
(g) Phantom section

Full Section

In full section the cutting plane passes entirely through an object. The resulting view will appear as illustrated. Although the plane usually passes along the main axis, it may be offset to reveal important features.

A full section showing the characteristic shape of an object usually replaces an exterior front view. One of the other principal views i.e., side or top view may be converted to a sectional view, however, if some interior feature can be shown to better advantage, or if the view is needed in addition to a sectioned front view.

The procedure in making a full section is simplified because it is an orthogonal view. The imaginary cut face is shown as it would appear to an observer looking directly at it an infinite distance away. In any sectional view, it is considered good practice to omit all hidden lines unless such lines are necessary to clarify the representation. Even then they should be used sparingly.

Half Section

The cutting plane for a half section removes one-quarter on an object. The plane cuts halfway through to the axis of the centre line so that half the finished sectional view appears in section and half appears as an external view.

This technique is used when a view is needed showing both the exterior and interior construction of a symmetrical object. Good practice dictates that hidden lines be omitted from both halves of the view unless they are absolutely necessary to explain the object.

The preferred method of representing the edge of the cutting plane in the sectional view is where the plane, being imaginary, is represented by the centre line of the object. An alternative method is to show the edge as a solid line.

Partial or Broken Section

A broken section is used mainly to expose the interior of objects where less than a half section is required for a satisfactory description. The object is theoretically cut, the front portion then removed by breaking material away.

The "breaking away" gives an irregular boundary to the section. No cutting plane edge is shown because the section is shown where it is broken.

Off-set Section

This is also known as section in two parallel planes. The cutting plane passes in a zig-zag way to reveal more clearly the inner details. The turning of the sectional plane gives a clear view. When an offset section is drawn, the edges of the offset section plane are not shown in sectional view, but the position or path of the offset section is shown in the related view by means of a cutting plane line.

Rotated or Revolved Section

A revolved section is useful for showing the true shape of the cross section of an object like a bar, arm, spoke or rib. This removes the need to draw a separate complete view.

To obtain a revolved section, pass an imaginary cutting plane through the arm, spoke or rib perpendicular to the longitudinal axis. Revolve the shape revealed by the cutting plane 90° into the plane of the paper. When revolved, the section should appear true shape and in its true revolved position. If any lines in the exterior view interfere with the revolved section they may be omitted. For better understanding, it may be necessary to break the object to provide an open space for the revolved section.

Removed (Detail) Section

A removed section is similar to a revolved section except that it does not appear where it was revolved but instead is drawn "out of place." A removed section appears adjacent to the view in which the section was taken. There are two reasons why removed sections are frequently desirable. First, their use may prevent a principal view with non uniform cross section from being cluttered with numerous revolved sections. Second, a section when removed may be drawn to an enlarged scale to emphasize detail and allow adequate dimensioning.

Whenever a removed section is used there must be some means of identifying it. Usually this is accomplished by showing the cutting plane on the principal view and by labelling both the plane and the resulting view.

A phantom section is a regular exterior view on which the interior construction is superimposed by crosshatching an imaginary cut surface with dashed section lines. This type of section is only used when a regular or broken section would remove some important detail, or in some instances, to show an accompanying part in relative position.

Section Lining

Section lines or hatching lines are thin uniformly spaced lines drawn usually at a 45^0 inclination to show the inner details distinctly. The spacing of these lines can be changed if sectioning has to be carried for more than two adjacent parts. Spacing between the lines should be uniform. The spacing depends upon the area to be sectioned. For small areas, the spacing is relatively less. The spacing gap can be taken from 1.5 mm to 3mm.

Thin Sections

Thin sections like, steel sections, gaskets, packing material, shims, etc., are shown entirely black. In steel joints a thin space may be left between two adjacent sections), the packing ring is shown entirely black in the sectional view of a pipe joint.

Machine Parts not Sectioned

Even when the section planes pass through the longitudinal axes of the machine parts such as shafts, handles, bolts, studs, screws, washers, nuts, rivets, keys, pins, gib, cotters, webs, stiffening ribs, lugs, spokes, arms, teeth of gears, bearings, etc., they are not hatched because they have no hidden details. Also, hatching might give a wrong idea of the depth or the thickness of the part. However, when such parts are cut across the axis, they should be shown in section.

Conventional Representation of Materials in Section

In view of the variety of materials used, it is desirable to represent different materials by conventions. The Bureau of Indian Standards has recommended the conventions to be used for the different materials. It has also been recommended that different materials should be indicated by notes on the drawing. For building materials, refer *IS: 962-1967*,"Code of practice for architectural and building drawings."

Assembly Sectioning

The assembly section is taken through a combination of parts of an object. It is necessary not only to show the cut surface but to distinguish between adjacent parts. This is done by drawing the section lines in different parts of the object. Section lines should be closer together in small parts.

It is necessary to change the inclination of the section lines in adjacent parts, so that they may be distinguished easily. In such a case where lines have the same inclination in the two adjacent parts the lines should not be in one alignment. The small amount of the clearance between mating or moving parts is not shown in the assembly drawing. If it is necessary to show, it should be shown at an enlarged scale, otherwise the clearance would have to be greatly exaggerated thus confusing the drawing. Even the clearance between a bolt and its hole, which may be as much as 1 mm is rarely shown. Machine elements such as bolts, nuts, shafts, etc. are not sectioned if they lie in the path of the cutting plane in the assembly drawing.

Principles of Section

Five principles of section are recommended and widely adopted in drafting practices. These principles are as follows:

(a) Nuts, bolts, screws, pins, rivets, shafts, rods, balls, rollers, keys, cotters, ribs spokes, arms, handles, wings and all other solid parts, whose axis occur in the plane of section are left in full and not sectioned.

(b) The cutting plane need not be a continuous, straight and single plane. It may be offset or bent at right angles at several places to pass through the unaligned features of an object. This helps to show the construction of the object in section to the best advantage.

(c) Invisible lines beyond the plane of section, particularly in full section, should not be drawn unless they are very necessary for description or identification of the object.

(d) Section lining, for the same piece, either in different view or in different parts of the same views, should be identical in spacing and direction.

(e) Adjacent parts of an object are shown by section lines drawn in opposite direction. These are often brought out more clearly by changing the alignment and varying the spacing of the section lines for each piece, using closer spacing for the smaller piece.

Technical Lettering

Technical lettering is the process of forming letters, numerals, and other characters in technical drawing. It is used to describe, or provide detailed specifications for, an object. With the goals of legibility and uniformity, styles are standardized and lettering ability has little relationship to normal writing ability. Engineering drawings use a Gothic sans-serif script, formed by a series of short strokes. Lower case letters are rare in most drawings of machines.

Methods of Forming Letters

1. *Freehand lettering* is done without the assistance of tools. To regulate lettering height, commonly 3 mm ([! in), guidelines are drawn.
2. *Mechanical lettering* is done using tools such as lettering guides, templates, or using a small mechanical pantograph referred to by the Keuffel and Esser trademark "Leroy"
3. Modern drawings are lettered with computer-aided design software.

Freehand Lettering

The letters to be drawn, though freehanded, should be stable and graceful. In some cases stability is impossible; for example, *P* and *F* are unavoidably top-heavy. In other cases the stability and grace of the letters may be maintained either by drawing the lower parts of the letters like *B*,*E* etc. wider than the upper parts, or by drawing the horizontal line at the centre of these letters just above their geometric axis. (Exception: In case of the letter *A*, the horizontal member is drawn below the geometric centre, to maintain *equality of areas below and above the centre line.* If a horizontal line is drawn exactly at the centre, then the difference in the areas of the triangle above the line and the trapezium below the line is much larger. This creates an unusual effect to our eyes.)

Emphasis should be on the overall beauty of a word, rather than individual letters. Most freehand lettering is done in a "gothic" style, i.e., with a constant line thickness; either "straight gothic", with vertical strokes perpendicular to the baseline, or "inclined gothic", with vertical strokes at about 75°.

Mechanical Lettering

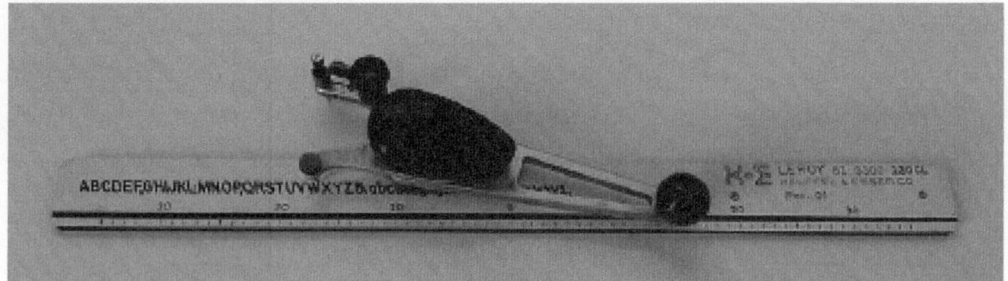

Figure: *K & E LEROY lettering set (1959)*

Mechanical lettering is sometimes done using a pantograph, a device consisting of four bars ("links") which are pinned to each other to form a parallelogram. The links can pivot about these pins. The lowermost link of the parallelogram is fixed to two rigid supports. One vertical link at one end is connected to a profile tracer, which traces the profile of the letter to be drawn, and the second vertical link and the other horizontal link are jointly connected to a pencil that draws the exact shape of the profile traced.

Dimensions of Letters

- *The Nominal Size of lettering* is defined by the height (h) of the outline contour of the upper-case (capital).
- *Central Line* is the imaginary line in the middle of each line or line element which is a constitutive part of a graphic character set.
- If we consider *d* as the width of the line element and *h* as the height of the line element, then the two standard ratios for d/h are: 1/14 and 1/10, which are feasible because they result in a minimum number of line thicknesses.

- *Location of Central Lines* – The nominal size (h) and the spacing between characters (a) shall be taken as the basis for defining the central line.
- *Range of Nominal Sizes* – The nominal size is typically one of the sequence 2.5 mm, 3.5 mm, 5 mm, 7 mm, 10 mm, 14 mm, 20 mm. Successive members of this sequence are approximately in a ratio of the square root of 2, as in the ISO 216 series of paper sizes.
- *Lettering Angle* – The lettering may be vertical (upright) or inclined (sloped) to the right at 75° from the horizontal.
- The spacing between two characters may be reduced by half, if this gives a better visual effect.
- Various letters are divided into number of parts so that dimensions will be accurate.
- The size of letter is described by its height. According to the height of letters, they are classified as Lettering A or Lettering B.

3

Technical Drawing

Technical drawing, also known as drafting or draughting, is the act and discipline of composing plans that visually communicate how something functions or has to be constructed. Drafting is the visual language of industry and engineering. A *drafter*, *draftsperson*, or *draughtsman* is a person who makes a drawing (technical or otherwise). A professional drafter who makes technical drawings is sometimes called a *drafting technician.*

People who communicate with technical drawings, (those who design and those who are tradespeople), may use technical standards that define practical symbols, perspectives, units of measurement, notation systems, visual styles, or layout conventions. These enable a drafter to communicate more concisely by using a commonly understood convention. Together, such conventions constitute a visual language, and help to ensure that the drawing is unambiguous and relatively easy to understand.

This need for unambiguous communication in the preparation of a functional document distinguishes technical drawing from the expressive drawing of the visual arts. Artistic drawings are subjectively interpreted; their meanings are multiply determined. Technical drawings are understood to have one intended meaning.

Methods

Sketching

A sketch is a quickly executed freehand drawing that is not intended as a finished work. In general, sketching is a quick way to record an idea for later use. Architect's sketches primarily serve as a way to try out different ideas and establish a composition before undertaking a more finished work, especially when the finished work is expensive and time consuming. Architectural sketches, for example, are a kind of diagrams. These sketches, like metaphors, are used by architects as a mean of communication in aiding design collaboration. This tool helps architects to abstract attributes of hypothetical provisional design solutions and summarize their complex patterns, hereby enhancing the design process.

Figure: Sketch for a government building.

Manual or by Instrument

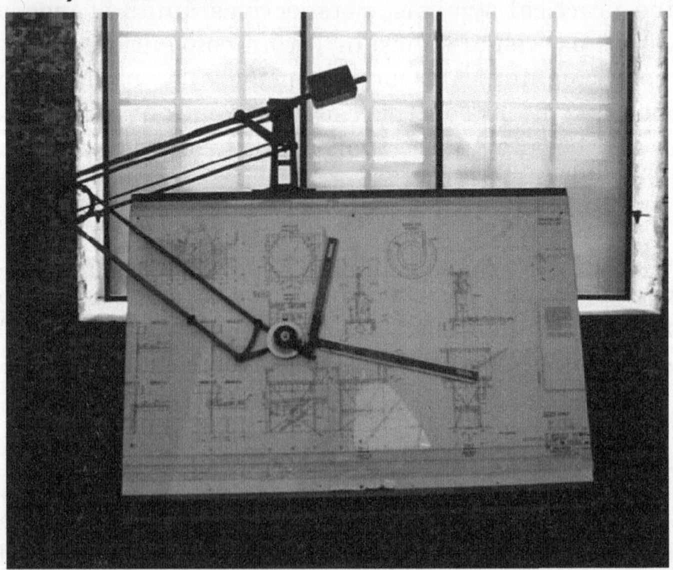

Figure: A drafting table.

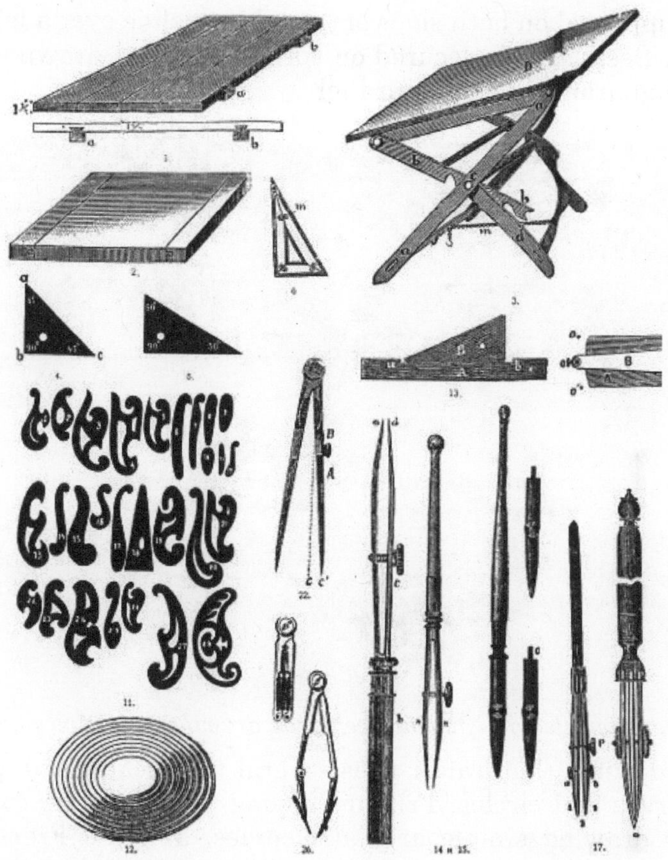

Figure: *Old-fashioned technical drawing instruments.*

The basic drafting procedure is to place a piece of paper (or other material) on a smooth surface with right-angle corners and straight sides—typically a drawing board. A sliding straightedge known as a *T-square* is then placed on one of the sides, allowing it to be slid across the side of the table, and over the surface of the paper.

"Parallel lines" can be drawn simply by moving the T-square and running a pencil or technical pen along the T-square's edge, but more typically the T-square is used as a tool to hold other devices such as set squares or triangles.

In this case the drafter places one or more triangles of known angles on the T-square—which is itself at right angles to the edge of the table—and can then draw lines at any chosen angle to others on the page. Modern drafting tables (which have by now largely been replaced by CAD workstations) come equipped with a drafting machine that is supported on both sides of the table to slide over a large piece of paper. Because it is secured on both sides, lines drawn along the edge are guaranteed to be parallel.

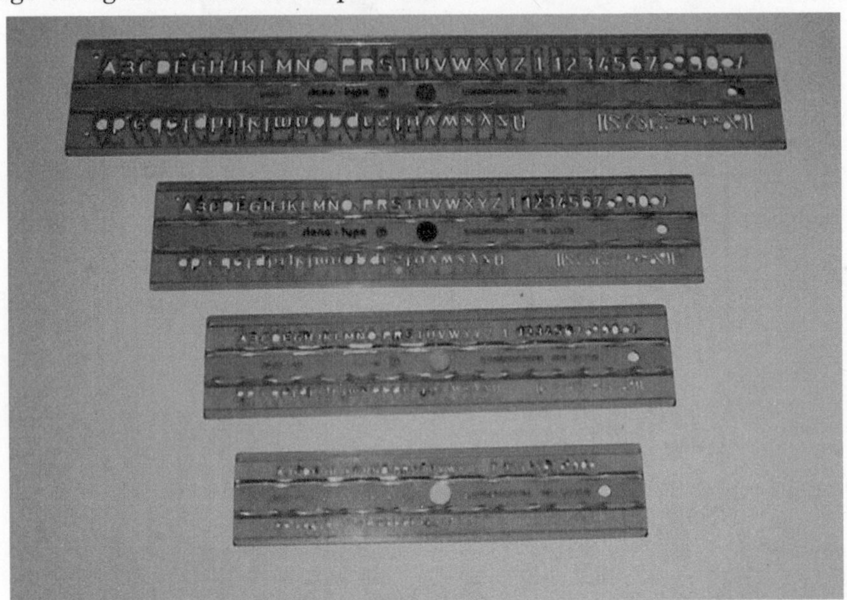

Figure: *Stencils for lettering technical drawings to DIN standards.*

In addition, the drafter uses several technical drawing tools to draw curves and circles. Primary among these are the compasses, used for drawing simple arcs and circles, and the French curve, typically a piece of plastic with complex curves on it. A spline is a rubber coated articulated metal that can be manually bent to most curves.

Drafting templates assist the drafter with creating recurring objects in a drawing without having to reproduce the object from scratch every time. This is especially useful when using common symbols; i.e. in the context of stagecraft, a lighting designer will typically draw from the USITT standard library of lighting fixture symbols to indicate the position of a common fixture across multiple positions. Templates are sold commercially by a number of vendors, usually customized to a specific task, but it is also not uncommon for a drafter to create their own templates.

This basic drafting system requires an accurate table and constant attention to the positioning of the tools. A common error is to allow the triangles to push the top of the T-square down slightly, thereby throwing off all angles. Even tasks as simple as drawing two angled lines meeting at a point require a number of moves of the T-square and triangles, and in general drafting can be a time consuming process. A solution to these problems was the introduction of the mechanical "drafting machine", an application of the pantograph (sometimes referred to incorrectly as a "pentagraph" in these situations) which allowed the drafter to have an accurate right angle at any point on the page quite quickly.

These machines often included the ability to change the angle, thereby removing the need for the triangles as well.

In addition to the mastery of the mechanics of drawing lines, arcs and circles (and text) onto a piece of paper—with respect to the detailing of physical objects—the drafting effort requires a thorough understanding of geometry, trigonometry and spatial comprehension, and in all cases demands precision and accuracy, and attention to detail of high order.

Although drafting is sometimes accomplished by a project engineer, architect—or even by shop personnel such as a machinist—skilled drafters (and/or designers) usually accomplish the task and are always in demand to some level.

***Figure:** A CAD model view of a four-cylinder inline crankshaft with pistons.*

Computer Aided Design

Today, the mechanics of the drafting task have largely been automated and accelerated through the use of computer-aided drawing systems (CAD).

There are two types of computer-aided design systems used for the production of technical drawings" two dimensions ("2D") and three dimensions ("3D").

2D CAD systems such as AutoCAD or MicroStation replace the paper drawing discipline. The lines, circles, arcs and curves are created within the software. It is down to the technical drawing skill of the user to produce the drawing. There is still much scope for error in the drawing when producing first and third angle orthographic projections, auxiliary projections and cross sections. A 2D CAD system is merely an electronic drawing board. Its greatest strength over direct to paper technical drawing is in the making of revisions. Whereas in a conventional hand drawn technical drawing, if a mistake is found, or a modification the is required, a new drawing must be made from scratch. The 2D CAD system allows a copy of the original to be modified, saving considerable time. 2D CAD systems can be used to create plans for large projects such as buildings and aircraft but provide no way to check the various components will fit together.

3D CAD systems such as Autodesk Inventor or SolidWorks first produce the geometry of the part, the technical drawing comes from user defined views of the part. Any orthographic, projected and section views are created by the software. There is no scope for error in the production of these views. The main scope for error comes in setting the parametre of first or third angle projection, and displaying the relevant symbol on the technical drawing. 3D CAD allows individual parts to be assembled together to represent the final product. Buildings, Aircraft, ships and cars are modelled, assembled and checked in 3D before technical drawings are released for manufacture.

Both 2D and 3D CAD systems can be used to produce technical drawings for any discipline. The various disciplines; electrical, electronic, pneumatic, hydraulic, etc., have industry recognised symbols to represent common components. BS and ISO produce standards to show recommended practices but it is up to individuals to produce the drawings. There is no definitive standard for layout or style. The only standard across engineering workshop drawings is in the creation of orthographic projections and cross section views.

Figure: *Architectural drawings typically begin with detailed measurements of the space which are then entered into a computer programme, as in this kitchen renovation project.*

Drafting can represent two dimensions ("2D") and three dimensions ("3D") although the representation itself is always created in 2D (cf. Architectural model). Drafting is the integral communication of technical or engineering drawings and is the industrial arts sub-discipline that underlies all involved technical endeavours.

In representing complex, three-dimensional objects in two-dimensional drawings, the objects can be described by at least one view plus material thickness note, 2, 3 or as many views and sections that are required to show all features of object.

Architectural Drawing

An architectural drawing or architect's drawing is a technical drawing of a building (or building project) that falls within the definition of architecture. Architectural drawings are used by architects and others for a number of purposes: to develop a design idea into a coherent proposal, to communicate ideas and concepts, to convince clients of the merits of a design, to enable a building contractor to construct it, as a record of the completed work, and to make a record of a building that already exists.

Architectural drawings are made according to a set of conventions, which include particular views (floor plan, section etc.), sheet sizes, units of measurement and scales, annotation and cross referencing. Conventionally, drawings were made in ink on paper or a similar material, and any copies required had to be laboriously made by hand. The twentieth century saw a shift to drawing on tracing paper, so that mechanical copies could be run off efficiently.

The development of the computer had a major impact on the methods used to design and create technical drawings, making manual drawing almost obsolete, and opening up new possibilities of form using organic shapes and complex geometry. Today the vast majority of drawings are created using CAD software.

Size and Scale

The size of drawings reflects the materials available and the size that is convenient to transport – rolled up or folded, laid out on a table, or pinned up on a wall. The draughting process may impose limitations on the size that is realistically workable. Sizes are determined by a consistent paper size system, according to local usage. Normally the largest paper size used in modern architectural practice is ISO A0 (841 mm × 1,189 mm or 33.1 in × 46.8 in) or in the USA Arch E (762 mm × 1,067 mm or 30 in × 42 in) or Large E size (915 mm × 1,220 mm or 36 in × 48 in).

Architectural drawings are drawn to scale, so that relative sizes are correctly represented. The scale is chosen both to ensure the whole building will fit on the chosen sheet size, and to show the required amount of detail. At the scale of one eighth of an inch to one foot (1:96) or the metric equivalent 1 to 100, walls are typically shown as simple outlines corresponding to the overall thickness. At a larger scale, half an inch to one foot (1:24) or the nearest common metric equivalent 1 to 20, the layers of different materials that make up the wall construction are shown. Construction details are drawn to a larger scale, in some cases full size (1 to 1 scale). Scale drawings enable dimensions to be 'read' off the drawing, i.e. measured directly. Imperial scales (feet and inches) are equally readable using an ordinary ruler.

On a one-eighth inch to one foot scale drawing, the one-eighth divisions on the ruler can be read off as feet. Architects normally use a scale ruler with different scales marked on each edge. A third method, used by builders in estimating, is to measure directly off the drawing and multiply by the scale factor.

Dimensions can be measured off drawings made on a stable medium such as vellum. All processes of reproduction introduce small errors, especially now that different copying methods mean that the same drawing may be re-copied or copies made in several different ways. Consequently dimensions need to be written on the drawing. The disclaimer "Do not scale off dimensions" is commonly inscribed on architects drawings, to guard against errors arising in the copying process.

Standard Views used in Architectural Drawing

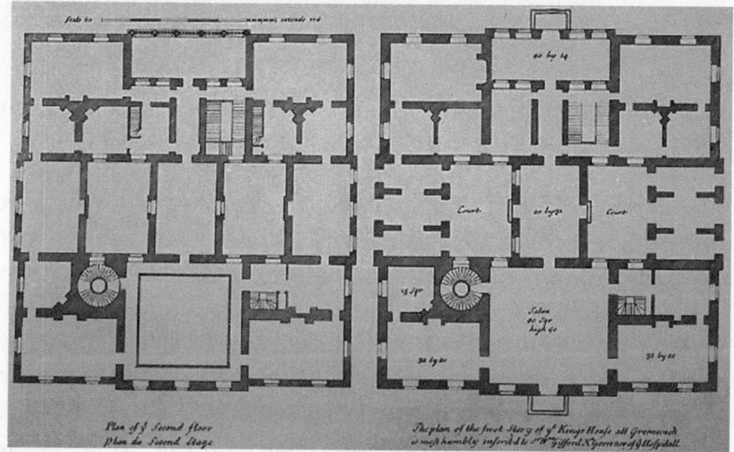

Figure: *Principal floor plans of the Queen's House, Greenwich (UK).*

Floor plan

A floor plan is the most fundamental architectural diagram, a view from above showing the arrangement of spaces in building in the same way as a map, but showing the arrangement at a particular level of a building. Technically it is a horizontal section cut through a building (conventionally at three feet / one metre above floor level), showing walls, windows and door openings and other features at that level. The plan view includes anything that could be seen below that level: the floor, stairs (but only up to the plan level), fittings and sometimes furniture. Objects above the plan level (e.g. beams overhead) can be indicated as dotted lines.

Geometrically, plan view is defined as a vertical orthographic projection of an object on to a horizontal plane, with the horizontal plane cutting through the building.

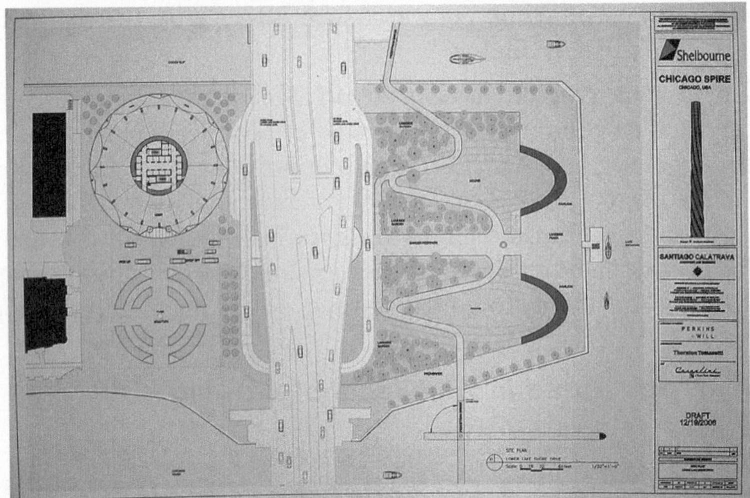

***Figure:** Site plan of the proposed Chicago Spire by Santiago Calatrava.*

Site plan

A site plan is a specific type of plan, showing the whole context of a building or group of buildings. A site plan shows property boundaries and means of access to the site, and nearby structures if they are relevant to the design. For a development on an urban site, the site plan may need to show adjoining streets to demonstrate how the design fits into the urban fabric. Within the site boundary, the site plan gives an overview of the entire scope of work. It shows the buildings (if any) already existing and those that are proposed, usually as a building footprint; roads, parking lots, footpaths, hard landscaping, trees and planting. For a construction project, the site plan also needs to show all the services connections: drainage and sewer lines, water supply, electrical and communications cables, exterior lighting etc.

Site plans are commonly used to represent a building proposal prior to detailed design: drawing up a site plan is a tool for deciding both the site layout and the size and orientation of proposed new buildings. A site plan is used to verify that a proposal complies with local development codes, including restrictions on historical sites. In this context the site plan forms part of a legal agreement, and there may be a requirement for it to be drawn up by a licenced professional: architect, engineer, landscape architect or land surveyor.

Figure: *Elevation of the principal façade of the Panthéon, Paris*

Elevation

An elevation is a view of a building seen from one side, a flat representation of one façade.

This is the most common view used to describe the external appearance of a building. Each elevation is labelled in relation to the compass direction it faces, e.g. the north elevation of a building is the side that most closely faces north.

Buildings are rarely a simple rectangular shape in plan, so a typical elevation may show all the parts of the building that are seen from a particular direction.

Geometrically, an elevation is a horizontal orthographic projection of a building on to a vertical plane, the vertical plane normally being parallel to one side of the building.

Architects also use the word elevation as a synonym for façade, so the north elevation is literally the north-facing wall of the building.

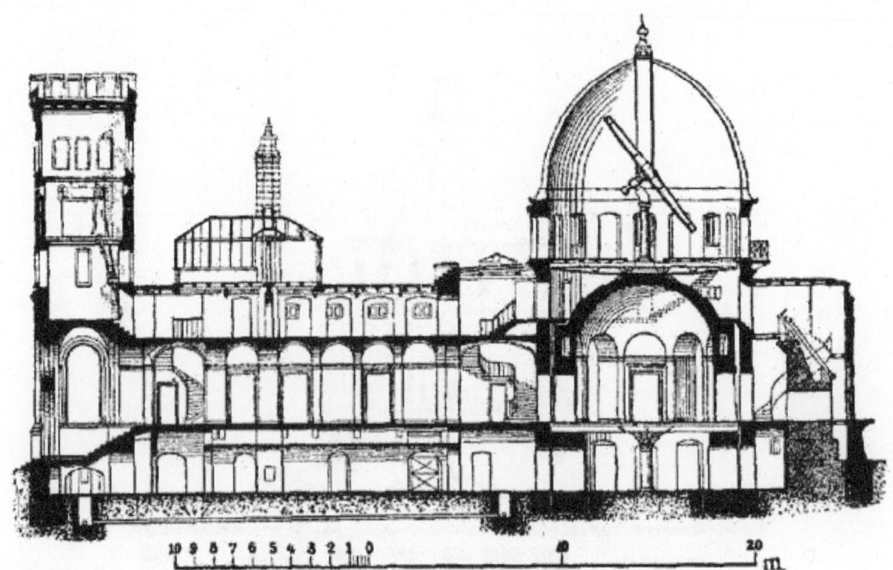

Aſtrophyſikaliſches Inſtitut zu Potsdam.

Figure: *Section drawing of the Observatorium at Potsdam.*

Cross Section

A cross section, also simply called a section, represents a vertical plane cut through the object, in the same way as a floor plan is a horizontal section viewed from the top.

In the section view, everything cut by the section plane is shown as a bold line, often with a solid fill to show objects that are cut through, and anything seen beyond generally shown in a thinner line.

Sections are used to describe the relationship between different levels of a building.

In the Observatorium drawing illustrated here, the section shows the dome seen from the outside, a second dome that can only be seen inside the building, and the way the space between the two accommodates a large astronomical telescope: relationships that would be difficult to understand from plans alone.

A sectional elevation is a combination of a cross section, with elevations of other parts of the building seen beyond the section plane.

Geometrically, a cross section is a horizontal orthographic projection of a building on to a vertical plane, with the vertical plane cutting through the building.

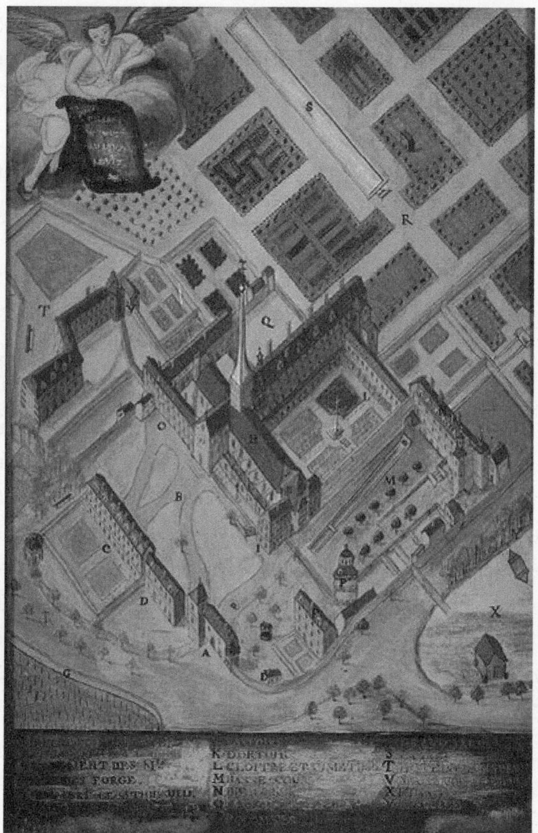

***Figure:** 18th century axonometric plan, Port-Royal-des-Champs.*

Isometric and Axonometric Projections

Isometric and axonometric projections are a simple way of representing a three dimensional object, keeping the elements to scale and showing the relationship between several sides of the same object, so that the complexities of a shape can be clearly understood.

There is some confusion about the terms isometric and axonometric. "Axonometric is a word that has been used by architects for hundreds of years. Engineers use the word axonometric as a generic term to include isometric, diametric and trimetric drawings." This article uses the terms in the architecture-specific sense.

Despite fairly complex geometrical explanations, for the purposes of practical draughting the difference between isometric and axonometric is simple. In both, the plan is drawn on a skewed or rotated grid, and the verticals are projected vertically on the page.

All lines are drawn to scale so that relationships between elements are accurate. In many cases a different scale is required for different axes, and again this can be calculated but in practice was often simply estimated by eye.

- An isometric uses a plan grid at 30 degrees from the horizontal in both directions, which distorts the plan shape. Isometric graph paper can be used to construct this kind of drawing. This view is useful to explain construction details (e.g. three dimensional joints in joinery). The isometric was the standard view until the mid twentieth century, remaining popular until the 1970s, especially for textbook diagrams and illustrations.
- Cabinet projection is similar, but only one axis is skewed, the others being horizontal and vertical. Originally used in cabinet making, the advantage is that a principal side (e.g. a cabinet front) is displayed without distortion, so only the less important sides are skewed. The lines leading away from the eye are drawn at a reduced scale to lessen the degree of distortion. The cabinet projection is seen in Victorian engraved advertisements and architectural textbooks, but has virtually disappeared from general use.
- An axonometric uses a 45 degree plan grid, which keeps the original orthogonal geometry of the plan. The great advantage of this view for architecture is that the draughtsman can work directly from a plan, without having to reconstruct it on a skewed grid. In theory the plan should be set at 45 degrees, but this introduces confusing coincidences where opposite corners align. Unwanted effects can be avoided by rotating the plan while still projecting vertically. This is sometimes called a planometric or plan oblique view, and allows freedom to choose any suitable angle to present the most useful view of an object.

Traditional draughting techniques used 30-60 and 45 degree set squares, and that determined the angles used in these views. Once the adjustable square became common those limitations were lifted.

The axonometric gained in popularity in the twentieth century, not just as a convenient diagram but as a formal presentation technique, adopted in particular by the Modern Movement. Axonometric drawings feature prominently in the influential 1970's drawings of Michael Graves, James Stirling and others, using not only straightforward views but worms-eye view, unusually and exaggerated rotations of the plan, and exploded elements. The axonometric view is not readily generated by CAD programmes which create views from a three dimensional model. Consequently it is now rarely used.

Detail Drawings

Detail drawings show a small part of the construction at a larger scale, to show how the component parts fit together. They are also used to show small surface details, for example decorative elements.

Section drawings at large scale are a standard way of showing building construction details, typically showing complex junctions (such as floor to wall junction, window openings, eaves and roof apex) that cannot be clearly shown on a drawing that includes the full height of the building.

A full set of construction details needs to show plan details as well as vertical section details. One detail is seldom produced in isolation: a set of details shows the information needed to understand the construction in three dimensions. Typical scales for details are 1/10, 1/5 and full size.

In traditional construction, many details were so fully standardised, that few detail drawings were required to construct a building. For example, the construction of a sash window would be left to the carpenter, who would fully understand what was required, but unique decorative details of the facade would be drawn up in detail. In contrast, modern buildings need to be fully detailed because of the proliferation of different products, methods and possible solutions.

Architectural Perspective

Perspective in drawing is an approximate representation on a flat surface of an image as it is perceived by the eye. The key concepts here are:

- Perspective is the view from a particular fixed viewpoint.
- Horizontal and vertical edges in the object are represented by horizontals and verticals in the drawing.
- Lines leading away into the distance appear to converge at a vanishing point.
- All horizontals converge to a point on the horizon, which is a horizontal line at eye level.
- Verticals converge to a point either above or below the horizon.

Figure: *Two point perspective, interior of Dercy House by Robert Adam, 1777.*

The basic categorization of artificial perspective is by the number of vanishing points:

- One-point perspective where objects facing the viewer are orthogonal, and receding lines converge to a single vanishing point.
- Two-point perspective reduces distortion by viewing objects at an angle, with all the horizontal lines receding to one of two vanishing points, both located on the horizon.
- Three-point perspective introduces additional realism by making the verticals recede to a third vanishing point, which is above or below depending upon whether the view is seen from above or below.

The normal convention in architectural perspective is to use two-point perspective, with all the verticals drawn as verticals on the page.

Three-point perspective gives a casual, photographic snapshot effect. In professional architectural photography, conversely, a view camera or a perspective control lens is used to eliminate the third vanishing point, so that all the verticals are vertical on the photograph, as with the perspective convention. This can also be done by digital manipulation of a photograph taken with a standard lens.

Aerial perspective is a technique in painting, for indicating distance by approximating the effect of the atmosphere on distant objects. In daylight, as an ordinary object gets further from the eye, its contrast with the background is reduced, its colour saturation is reduced, and its colour becomes more blue.

Not to be confused with aerial view or bird's eye view, which is the view as seen (or imagined) from a high vantage point. In J M Gandy's perspective of the Bank of England, Gandy portrayed the building as a picturesque ruin in order to show the internal plan arrangement, a precursor of the cutaway view.

A montage image is produced by superimposing a perspective image of a building on to a photographic background. Care is needed to record the position from which the photograph was taken, and to generate the perspective using the same viewpoint. This technique is popular in computer visualisation, where the building can be photorealistically rendered, and the final image is intended to be almost indistinguishable from a photograph.

Sketches and Diagrams

Figure: *Architect's early concept sketches.*

A sketch is a rapidly executed freehand drawing, a quick way to record and develop an idea, not intended as a finished work. A diagram may also be drawn freehand but deals with symbols, to develop the logic of a design. Both may be worked up into a more presentable form and used to communicate the principles of a design.

In architecture, the finished work is expensive and time consuming, so it is important to resolve the design as fully as possible before construction work begins. Complex modern buildings involve a large team of different specialist disciplines, and communication at the early design stages is essential to keep the design moving towards a coordinated outcome. Architects (and other designers) start investigating a new design with sketches and diagrams, to develop a rough design that provides an adequate response to the particular design problems. There are two basic elements to a building design, the aesthetic and the practical. The aesthetic element includes the layout and visual appearance, the anticipated feel of the materials, and cultural references that will influence the way people perceive the building. Practical concerns include space allocated for different activities, how people enter and move around the building, daylight and artificial lighting, acoustics, traffic noise, legal matters and building codes, and many other issues. While both aspects are partly a matter of customary practice, every site is different. Many architects actively seek innovation, thereby increasing the number of problems to be resolved.

Architectural legend often refers to designs made on the back of an envelope/napkin/cigarette packet. Initial thoughts are important, even if they have to be discarded along the way, because they provide the central idea around which the design can develop. Although a sketch is inaccurate, it is disposable and allows for freedom of thought, for trying different ideas quickly. Choice becomes sharply reduced once the design is committed to a scale drawing, and the sketch stage is almost always essential. Diagrams are mainly used to resolve practical matters. In the early phases of the design architects use diagrams to develop, explore, and communicate ideas and solutions. They are essential tools for thinking, problem solving, and communication in the design disciplines. Diagrams can be used to resolve spatial relationships, but they can also represent forces and flows, e.g. the forces of sun and wind, or the flows of people and materials through a building.

An exploded view diagram shows component parts dis-assembled in some way, so that each can be seen on its own. These views are common in technical manuals, but are also used in architecture, either in conceptual diagrams or to illustrate technical details. In a cutaway view parts of the exterior are omitted to show the interior, or details of internal construction. Although common in technical illustration, including many building products and systems, the cutaway is in fact little-used in architectural drawing.

Types of Architectural Drawing

Architectural drawings are produced for a specific purpose, and can be classified accordingly. Several elements are often included on the same sheet, for example a sheet showing a plan together with the principal façade.

Presentation Drawings

Drawings intended to explain a scheme and to promote its merits. Working drawings may include tones or hatches to emphasise different materials, but they are diagrams, not intended to appear realistic. Basic presentation drawings typically include people, vehicles and trees, taken from a library of such images, and are otherwise very similar in style to working drawings.

Rendering is the art of adding surface textures and shadows to show the visual qualities of a building more realistically. An architectural illustrator or graphic designer may be employed to prepare specialist presentation images, usually perspectives or highly finished site plans, floor plans and elevations etc.

Survey Drawings

Measured drawings of existing land, structures and buildings. Architects need an accurate set of survey drawings as a basis for their working drawings, to establish exact dimensions for the construction work. Surveys are usually measured and drawn up by specialist land surveyors.

Record Drawings

Historically, architects have made record drawings in order to understand and emulate the great architecture known to them. In the Renaissance, architects from all over Europe studied and recorded the remains of the Roman and Greek civilizations, and used these influences to develop the architecture of the period.

Records are made both individually, for local purposes, and on a large scale for publication. Historic surveys worth referring to include:

- Colen Campbell's *Vitruvius Brittanicus*, illustrations of English buildings by Inigo Jones and Sir Christopher Wren, as well as Campbell himself and other prominent architects of the era.
- The Survey of London, founded in 1894 by Charles Robert Ashbee and now available through English Heritage. A record of notable streets and individual buildings in the former County of London.
- Historic American Buildings Survey, records of notable buildings drawn up during the 1930s Depression, this collection is held by the Library of Congress and is available copyright-free on the internet.

Record drawings are also used in construction projects, where "as-built" drawings of the completed building take account of all the variations made during the course of construction.

Working Drawings

A comprehensive set of drawings used in a building construction project: these will include not only architect's drawings but structural and services engineer's drawings etc. Working drawings logically subdivide into location, assembly and component drawings.

- Location drawings, also called general arrangement drawings, include floor plans, sections and elevations: they show where the construction elements are located.
- Assembly drawings show how the different parts are put together. For example a wall detail will show the layers that make up the construction, how they are fixed to structural elements, how to finish the edges of openings, and how prefabricated components are to be fitted.
- Component drawings enable self-contained elements e.g. windows and doorsets, to be fabricated in a workshop, and delivered to site complete and ready for installation. Larger components may include roof trusses, cladding panels, cupboards and kitchens. Complete rooms, especially hotel bedrooms and bathrooms, may be made as prefabricated pods complete with internal decorations and fittings.

Traditionally, working drawings would typically combine plans, sections, elevations and some details to provide a complete explanation of a building on one sheet. That was possible because little detail was included, the building techniques involved being common knowledge amongst building professionals. Modern working drawings are much more detailed and it is standard practice to isolate each view on a separate sheet. Notes included on drawings are brief, referring to standardised specification documents for more information. Understanding the layout and construction of a modern building involves studying an often-sizeable set of drawings and documents.

Drafting

Until the latter part of the twentieth century, all architectural drawings were manually produced, either by architects or by trained (but less skilled) draughtsmen (or drafters), who did not generate the design, although they made many of the less important decisions. This system continues with CAD draughting: many design architects have little or no knowledge of CAD software programmes and rely upon others to take their designs beyond the sketch stage. Draughtsmen may specialize in a type of structure, such as residential or commercial, or in a type of construction: timber frame, reinforced concrete, prefabrication etc.

Figure: *Architect at his drawing board, 1893*

The traditional tools of the architect were the drawing board or draughting table, T-square and set squares, protractor, compasses, pencil and drawing pens of different types. Drawings were made on vellum, coated linen, and on tracing paper.

Lettering would either be done by hand, mechanically using a stencil, or a combination of the two. Ink lines were drawn with a ruling pen, a relatively sophisticated device similar to a dip-in pen but with adjustable line width, capable of producing a very fine controlled line width. Ink pens had to be dipped into ink frequently. Draughtsmen worked standing up, and kept the ink on a separate table to avoid spilling ink on the drawing.

Twentieth century developments include the parallel motion drawing board, and more complicated improvements on the basic T-square. The development of reliable technical drawing pens allowed for faster draughting and stencilled lettering. Letraset dry transfer lettering and half-tone sheets were popular from the 1970s until computers made those processes obsolete.

Computer-aided Design

***Figure:** Computer generated perspective of the Moscow School of Management, by David Adjaye*

Computer-aided design is the use of computer software to create drawings. Today the vast majority of technical drawings of all kinds are made using CAD. Instead of drawing lines on paper, the computer records equivalent information electronically. There are many advantages to this system: repetition is reduced because complex elements can be copied, duplicated and stored for re-use. Errors can be deleted, and the speed of draughting allows many permutations to be tried before the design is finalised. On the other hand, CAD drawing encourages a proliferation of detail and increased expectations of accuracy, aspects which reduce the efficiency originally expected from the move to computerisation.

Professional CAD software such as AutoCAD is complex and requires both training and experience before the operator becomes fully productive. Consequently, skilled CAD operators are often divorced from the design process. Simpler software such as SketchUp allows for more intuitive drawing and is intended as a design tool.

CAD is used to create all kinds of drawings, from working drawings to photorealistic perspective views. Architectural renderings (also called visualisations) are made by creating a three-dimensional model using CAD. The model can be viewed from any direction to find the most useful viewpoints. Different software (for example Autodesk 3ds Max) is then used to apply colour and texture to surfaces, and to represent shadows and reflections. The result can be accurately combined with photographic elements: people, cars, background landscape.

Building information modelling (BIM) is the logical development of CAD drawing, a relatively new technology but fast becoming mainstream. The design team collaborates to create a three-dimensional computer model, and all plans and other two-dimensional views are generated directly from the model, ensuring spatial consistency. The key innovation here is to share the model via the internet, so that all the design functions (site survey, architecture, structure and services) can be integrated into a single model, or as a series of models associated with each specialism that are shared throughout the design development process. Some form of management, not necessarily by the architect, needs to be in place to resolve conflicting priorities. The starting point of BIM is spatial design, but it also enables components to be quantified and scheduled directly from the information embedded in the model.

An architectural animation is a short film showing how a proposed building will look: the moving image makes three-dimensional forms much easier to understand. An animation is generated from a series of hundreds or even thousands of still images, each made in the same way as an architectural visualisation. A computer-generated building is created using a CAD programme, and that is used to create more or less realistic views from a sequence of viewpoints. The simplest animations use a moving viewpoint, while more complex animations can include moving objects: people, vehicles and so on.

Architectural Reprography

Architectural reprography, the reprography of architectural drawings, covers a variety of technologies, media, and supports typically used to make multiple copies of original technical drawings and related records created by architects, landscape architects, engineers, surveyors, mapmakers and other professionals in building and engineering trades.

Within the context of archival preservation, the custodians of architectural records must consider many aspects of identification and care when managing the artifactual nature of these materials.

Storage containers, handling, paper and chemical compositions and interactions, ultraviolet light exposure, humidity, mold, and other agents of potential harm all interact to determine the longevity of these documents. As well, architectural reprographic drawings are often in very large formats, making storage and handling decisions especially complex.

History

With the rise of the professionalized practice of western architecture in the second half of the 19th century, the field of architectural reprography—and the corresponding developments of photography and mass-produced wood-pulp paper—saw significant experiments and advances in technology.

Beginning with major refinements in blueprinting processes in the 1840s, through the widespread adoption of diazotype printing after World War II, the design profession turned to analog architectural reprography to create accurate, to-scale reproductions of original drawings created on tracing paper, vellum, and linen supports.

These copies were typically used throughout the architect's own design process and also for distribution to clients, contractors, governmental agencies, and other interested parties.

However, the integration of CAD—or Computer-Aided Design—over the last twenty-five years of design practice has made analog reprography far less common in the profession and more ephemeral in nature.

For archivists, curators, librarians and other custodians of architectural records, traditional reprographic formats are now often seen as historic documents, with attendant needs for long-term care and conservation.

Major Processes

Both the underlying support—paper or plastic—and the image type are used to identify the specific processes used in architectural reprography. Between the late 19th century and the late 20th century, several processes emerged as the preferred methods, used for decades, while other less common processes were employed for shorter periods of time.

Blueprints

First developed in 1725, blueprinting uses a wet process to produce an image of white lines on a cyan or Prussian blue ground. To make a blueprint, a heavy paper (or more rarely drafting linen) support is impregnated with potassium ferricyanide and ferric ammonium, placed under a translucent original drawing, weighted with glass, and exposed to ultraviolet light. After sufficient light exposure, the glass and original drawing are removed and the blueprint paper is washed to reveal a negative image. This same process, using an intermediary reprographic drawing, could also be used to produce a positive blueprint—blue lines on a white ground—however, this more expensive and time-intensive method was far less commonly employed.

The major disadvantages of the blueprint process, however, included paper distortions caused by the wet process which might render scale drawings less accurately, as well as the inability to make further copies from the blueprints. Nonetheless, for its efficiency and low cost, the blueprint process, further simplified and mechanized by the turn of the 20th century, became the most widely-used reprographic process between the mid-19th century and the latter half of the 20th-century.

In archival settings, because the process involves ammonium, the resulting prints should not be stored in contact with other papers that have a buffered reserve, nor should blueprints be de-acidified, as the resulting chemical interactions can cause irreversible image loss. Blueprints are also highly light-sensitive and should not be exposed to ultraviolet light for long periods of time.

Pellet Prints

Invented in 1887 by Henry Pellet, the Pellet process uses a wet process to produce an image of cyan or Prussian blue lines on a white ground. Essentially, this process produces a positive image, while a blueprint produces a negative one. To make a Pellet print, a paper (or more rarely drafting linen) support is coated with ferric salts suspended in a gelatin emulsion, placed under a translucent original drawing, weighted with glass, and exposed to ultraviolet light. As with the blueprint process, after sufficient light exposure, the original drawing is removed, the paper washed in a ferrocyanide bath, and then rinsed in an acidic bath to reveal a positive image. This process required fewer steps than creating a positive blueprint, and was thus more widely employed during the late 19th and early 20th centuries.

In an archival setting, Pellet prints should be treated and stored under the same conditions as blueprints.

Van Dyke Prints

The Van Dyke process, invented by F. R. Van Dyke in 1901, created an intermediary print—a white line on a dark brown ground—that could used in any of several other processes, such as blueprinting, to create a positive print, i.e. a dark line on a light ground. Using a translucent vellum support, the paper was prepared with a coating of silver salts. The vellum was then united with the original drawing, exposed to ultraviolet light, and later washed in a sodium thiosulphate bath.

In an archival setting, Van Dyke prints are relatively rare, as they were created for temporary purposes and often discarded after the final positive prints were made. Because of the nitrates used in preparing the paper and the preferred thin paper itself, Van Dyke prints are often extremely brittle and susceptible to damage. Van Dyke prints should be stored separately and, when possible reformatted before the image degrades unacceptably.

Diazotypes

By the middle of the 20th century, wet-process reprographic techniques such as blueprinting, Pellet, and Van Dyke printing were largely superseded by various dry-printing processes. The most common of these is the Diazotype process, refined in 1920s, which used paper supports sensitized with diazonium salts, a coupling agent, and an acid stabilizer to produce a dark line on a white ground. The Diazo positive print was considered more readable than a negative blueprint, and the dry process eliminated the image distortion of wet paper.

As with other earlier reprographic processes, a translucent original drawing was placed over a sheet of the sensitized paper and exposed to light. However, the next step exposed the paper to an ammonia gas. This alkaline gas catalyzed a reaction between the diazo salts and the coupling agent to produce an image that fixed in the paper over several days. Typically these prints have blue or dark purple lines on a mottled cream-coloured background, although line and ground colours can vary.

A related process is the sepia Diazo print, which produced either a positive or negative print in dark brown and light tones. The negative versions of these prints were most often produced as intermediaries, like the earlier Van Dyke process, to allow corrections and revisions without disturbing the original drawing. In the negative printing process, additional resins and oils were sometimes added to the paper support to increase translucency. Positive sepia prints, generally made on opaque paper, were typically used as an alternative to positive blueline Diazo prints. Both blueline and sepia prints were often poorly and cheaply processed, resulting in undesirable residual chemical content. Off-gassing of sulphurous compounds, image fading, and yellowing of the paper support are common signs of degradation and are not reversible. Diazo prints are also highly light-sensitive and can fade to illegibility within a short period of exposure to ultraviolet light.

In archival practice, Diazo prints are the most common reprographic format encountered in late 20th-century architectural collections. However, their inherent fragility and fugitive images, as compared with blueprints and earlier processes, makes their care problematic. Diazos—particularly sepia prints, which readily transfer color to adjacent papers—should be physically segregated from all other types of media. Exposure to light and pollutants in air should be minimized, and wherever possible, original drawings or reformatted prints should be kept for reference.

Other Processes

- Hectographic prints
- Ferrogallic prints
- Gel-lithographs
- Photostatic prints
- Wash-Off prints
- Silver halide prints
- Electrostatic prints

Cleaning, Flattening, and Repairing

For large collections of architectural materials, conservation work can address several areas of concern. Consultation with a professional conservator is recommended, although some minor treatments can be accomplished by general caretakers with training. Rolled and folded reprography, once cleaned, can be flattened through humidification. Cleaning may be done with white vinyl erasers, using great care in areas of friable media, such as graphite and coloured pencil. Tears, losses, and other surface damage should be treated by a professional conservator. For particularly fragile or frequently-handled prints, sheets may be encapsulated in polyester or polypropylene film for additional support and protection. This is not recommended, however, for reprographic prints with annotations in friable media.

Storage

Rolled Storage

The most common form of storage for architectural drawings—both for drawings in active professional use and in archival environments—has traditionally been in rolls. While this allows for efficiency in the use of space and ease of retrieval, potentially damaging situations can arise from a casual approach to roll storage. For reprographic drawings on paper supports, rolling can stress paper fibres and make unrolling for examination more difficult. Small rolls can be easily crushed and ends can be creased and torn without additional protective wrapping and support.

Flat Storage

In circumstances where fragile, rigid, or otherwise atypical media makes rolled storage unfeasible, storage in flat boxes or flatfile drawers can be the best choice. Acid-free and lignin-free portfolio boxes, ideally no more than four inches deep, can be cost-effective and allow more flexibility in arrangement on shelving. Flatfile furniture should meet the minimum requirements of archivally-sound construction—powder- or enamel-coated steel units with no rust or sharp edges that could damage materials while stored or moved in and out of the drawers.

Drawings should be grouped and identified for ease in retrieval, preferably within folders that are cut to fit the full dimensions of the corresponding container. As with rolled materials, the potentially damaging chemical interactions of print processes should be considered when grouping drawings in folders. Wherever possible, for example, blueprints should be segregated from diazotypes, and sepia diazo prints should be stored alone to the extent possible.

Reformatting

For most drawings, especially those that are oversized or significantly damaged, photographic reproduction remains the best method of accurately reproducing the fine details of a drawing. For drawings that are not significantly damaged or that are encapsulated in a polyester film, digital flat-bed scanning or other mechanical methods may be used.

Professional Resources

The Society of American Archivists supports many architectural archivists in their professional responsibilities. In particular, the SAA's Architectural Records Roundtable is a primary forum for discussion of issues of acquisition, identification, description, conservation, and digital preservation of a wide variety of architectural documentation.

Dimensioning

Dimensioning is the process of measuring the cubic space that a package or object occupies. It is the method of calculating dimensional weight for the storage, handling, transporting and invoicing of goods. Vehicles and storage units have both volume and weight capacity limits and can easily become full in terms of volume before they reach their capacity in weight. By dimensioning objects, parcels and pallets, shipping companies and warehouses can make optimal use of space and charge for services accordingly.

The Background Behind Dimensioning

In 1985 a Norwegian company called Cargoscan, which officially became part of the Mettler Toledo group in 2000, began supplying dimensioners and data capture solutions to transport and logistics companies around the world. The aim was to provide the industry with a system for revenue recovery and protection. It seemed that if carriers were to take both weight and volume into consideration when invoicing, they could charge customers more fairly according to services provided and make better use of resources. 0

In Transport and Logistics

Shipping costs have historically been calculated based on weight alone. By charging only by weight, lightweight, low density packages become unprofitable for freight carriers due to the amount of space they take up in trucks/aircraft/ships in proportion to their actual weight. By using dimensioning technology to calculate an item's dimensional weight, carriers are able to charge based on either volume or weight, depending on which is greater.

In the warehousing industry, dimensioning is used to provide an overview of the volume items in stock which can reduce the costs of materials, return handling, shipping and manpower.

The main reasons that companies choose to invest in automated dimensioning systems include revenue recovery, increased throughput, reduced returns and quality control.

Dimensioning Systems are not only used by the biggest companies in the industry such as DHL, UPS, FedEx and TNT but also by postal companies, shipping retailers, and small handling companies around the world.

Locus (Mathematics)

In Geometry, a locus (plural: *loci*) is a set of points whose location satisfies or is determined by one or more specified conditions i.e., 1)every point satisfies a given condition and 2)every point satisfiying it is in that particular locus

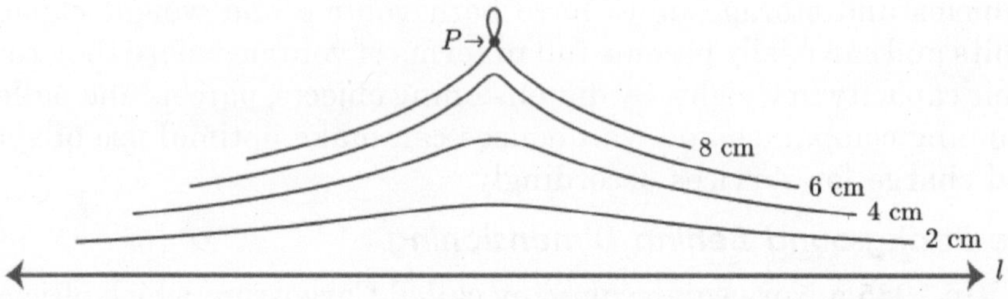

***Figure:** Each curve in this example is the* locus *of a set of points that lie on any line defined as the conchoid of a circle centred at point* P *and the line* l. *In this example,* P *is 7cm from* l.

Commonly Studied Loci

Examples from plane geometry:

- The set of points equidistant from two points is a perpendicular bisector to the line segment connecting the two points.
- The set of points equidistant from two lines which cross is the angle bisector.
- All conic sections are loci:
  - o Parabola: the set of points equidistant from a single point (the focus) and a line (the directrix).
  - o Circle: the set of points for which the distance from a single point is constant (the radius). The set of points for each of which the ratio of the distances to two given foci is a positive constant (that is not 1) is referred to as a Circle of Apollonius.
  - o Hyperbola: the set of points for each of which the absolute value of the difference between the distances to two given foci is a constant.

- o Ellipse: the set of points for each of which the sum of the distances to two given foci is a constant. In particular, the circle is a locus.

Proof of a Locus

In order to prove the correctness of a locus, one generally divides the proof into two stages:

- Proof that all the points that satisfy the conditions are on the locus.
- Proof that all the points on the locus satisfy the conditions.

Example

$$\sqrt{(x+1)^2+(y-0)^2}=3\sqrt{(x-0)^2+(y-2)^2}$$

On squaring,

$$(x+1)^2+(y-0)^2=9(x-0)^2+9(y-2)^2 \Leftrightarrow 8(x^2+y^2)-2x-36y+35=0$$

This is equivalent to

$$\left(x-\frac{1}{8}\right)^2+\left(y-\frac{9}{4}\right)^2=\frac{45}{64}$$

This equation represents a circle.

Conic section

In mathematics, a conic section (or just conic) is a curve obtained as the intersection of a cone (more precisely, a right circular conical surface) with a plane. In analytic geometry, a conic may be defined as a plane algebraic curve of degree 2. There are a number of other geometric definitions possible. One of the most useful, in that it involves only the plane, is that a conic consists of those points whose distances to some point, called a *focus*, and some line, called a *directrix*, are in a fixed ratio, called the *eccentricity*.

Traditionally, the three types of conic section are the hyperbola, the parabola, and the ellipse. The circle is a special case of the ellipse, and is of sufficient interest in its own right that it is sometimes called the fourth type of conic section.

The type of a conic corresponds to its eccentricity, those with eccentricity less than 1 being ellipses, those with eccentricity equal to 1 being parabolas, and those with eccentricity greater than 1 being hyperbolas. In the focus-directrix definition of a conic the circle is a limiting case with eccentricity 0. In modern geometry certain degenerate cases, such as the union of two lines, are included as conics as well. The conic sections were named and studied at least since 200 BC, when Apollonius of Perga undertook a systematic study of their properties.

History

Menaechmus and Early Works

It is believed that the first definition of a conic section is due to Menaechmus (died 320 BC). His work did not survive and is only known through secondary accounts.

The definition used at that time differs from the one commonly used today in that it requires the plane cutting the cone to be perpendicular to one of the lines, (a generatrix), that generates the cone as a surface of revolution.

Thus the shape of the conic is determined by the angle formed at the vertex of the cone (between two opposite generatrices): If the angle is acute then the conic is an ellipse; if the angle is right then the conic is a parabola; and if the angle is obtuse then the conic is a hyperbola. Note that the circle cannot be defined this way and was not considered a conic at this time.

Euclid (fl. 300 BC) is said to have written four books on conics but these were lost as well. Archimedes (died c. 212 BC) is known to have studied conics, having determined the area bounded by a parabola and an ellipse. The only part of this work to survive is a book on the solids of revolution of conics.

Apollonius of Perga

The greatest progress in the study of conics by the ancient Greeks is due to Apollonius of Perga (died c.190 BC)), whose eight volume *Conic Sections* summarized the existing knowledge at the time and greatly extended it.

Apollonius's major innovation was to characterize a conic using properties within the plane and intrinsic to the curve; this greatly simplified analysis. With this tool, it was now possible to show that any plane cutting the cone, regardless of its angle, will produce a conic according to the earlier definition, leading to the definition commonly used today. Pappus of Alexandria (died c. 350 CE) is credited with discovering importance of the concept of a focus of a conic, and the discovery of the related concept of a directrix.

Al-Kuhi

An instrument for drawing conic sections was first described in 1000 CE by the Islamic mathematician Al-Kuhi.

Omar Khayyám

Apollonius's work was translated into Arabic (the technical language of the time) and much of his work only survives through the Arabic version. Persians found applications to the theory; the most notable of these was the Persian mathematician and poet Omar Khayyám who used conic sections to solve algebraic equations.

Europe

Johannes Kepler extended the theory of conics through the "principle of continuity", a precursor to the concept of limits. Girard Desargues and Blaise Pascal developed a theory of conics using an early form of projective geometry and this helped to provide impetus for the study of this new field. In particular, Pascal discovered a theorem known as the hexagrammum mysticum from which many other properties of conics can be deduced. Meanwhile, René Descartes applied his newly discovered Analytic geometry to the study of conics. This had the effect of reducing the geometrical problems of conics to problems in algebra.

Features

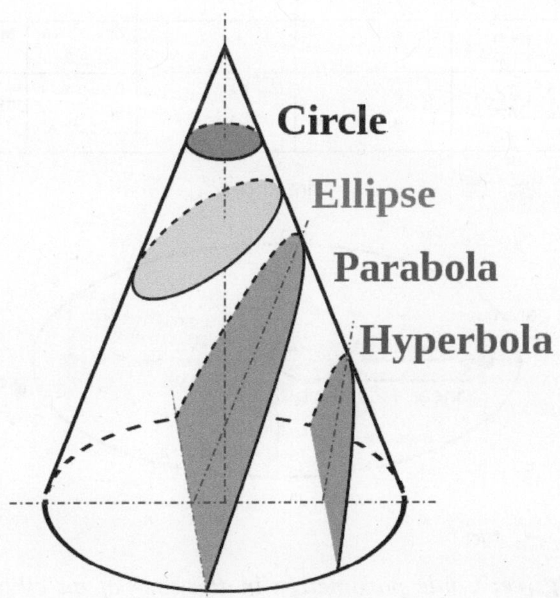

The three types of conics are the ellipse, parabola, and hyperbola. The circle can be considered as a fourth type (as it was by Apollonius) or as a kind of ellipse. The circle and the ellipse arise when the intersection of cone and plane is a closed curve. The circle is obtained when the cutting plane is parallel to the plane of the generating circle of the cone – for a right cone as in the picture at the top of the page this means that the cutting plane is perpendicular to the symmetry axis of the cone. If the cutting plane is parallel to exactly one generating line of the cone, then the conic is unbounded and is called a parabola. In the remaining case, the figure is a hyperbola. In this case, the plane will intersect *both* halves (*nappes*) of the cone, producing two separate unbounded curves.

Various parametres are associated with a conic section, as shown in the following table. (For the ellipse, the table gives the case of $a>b$, for which the major axis is horizontal; for the reverse case, interchange the symbols a and b. For the hyperbola the east-west opening case is given. In all cases, a and b are positive.)

CONIC SECTION	*EQUATION*	*ECCENTRICITY (E)*	*LINEAR ECCENTRICITY (C)*	*SEMI-LATUS RECTUM (ℓ)*	*FOCAL PARAMETER (P)*
circle	$x^2+y^2=a^2$	0	0	a	∞
ellipse	$\frac{x^2}{a^2}+\frac{y^2}{b^2}=1$	$\sqrt{1-\frac{b^2}{a^2}}$	$\sqrt{a^2-b^2}$	$\frac{b^2}{a}$	$\frac{b^2}{\sqrt{a^2-b^2}}$
parabola	$y^2=4ax$	1	a	2a	2a
hyperbola	$\frac{x^2}{a^2}-\frac{y^2}{b^2}=1$	$\sqrt{1+\frac{b^2}{a^2}}$	$\sqrt{a^2+b^2}$	$\frac{b^2}{a}$	$\frac{b^2}{\sqrt{a^2+b^2}}$

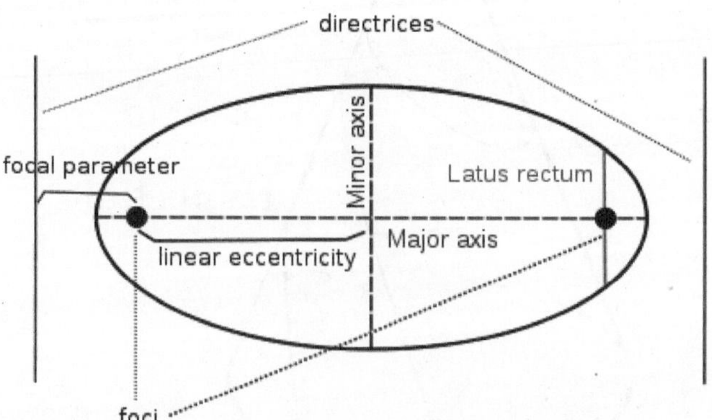

Figure: *Conic parametres in the case of an ellipse*

Conic sections are exactly those curves that, for a point F, a line L not containing F and a non-negative number e, are the locus of points whose distance to F equals e times their distance to L. F is called the focus, L the directrix, and e the eccentricity.

The linear eccentricity (c) is the distance between the centre and the focus (or one of the two foci).

The latus rectum (2ℓ) is the chord parallel to the directrix and passing through the focus (or one of the two foci).

The semi-latus rectum (ℓ) is half the latus rectum.

The focal parametre (p) is the distance from the focus (or one of the two foci) to the directrix.

The following relations hold:

- $pe = \ell$
- $ae = c$.

Construction

There are many methods to construct a conic. One of them, that is useful in engineering applications, being parallelogram method, where a conic is constructed point by point by means of connecting certain equally spaced points on horizontal line and vertical line.

Properties

Just as two (distinct) points determine a line, five points determine a conic. Formally, given any five points in the plane in general linear position, meaning no three collinear, there is a unique conic passing through them, which will be non-degenerate; this is true over both the affine plane and projective plane. Indeed, given any five points there is a conic passing through them, but if three of the points are collinear the conic will be degenerate (reducible, because it contains a line), and may not be unique. Irreducible conic sections are always "smooth". This is important for many applications, such as aerodynamics, where a smooth surface is required to ensure laminar flow and to prevent turbulence.

Intersection at Infinity

An algebro-geometrically intrinsic form of this classification is by the intersection of the conic with the line at infinity, which gives further insight into their geometry:

- ellipses intersect the line at infinity in 0 points—rather, in 0 real points, but in 2 complex points, which are conjugate;

- parabolas intersect the line at infinity in 1 double point, corresponding to the axis—they are tangent to the line at infinity, and close at infinity, as distended ellipses;
- hyperbolas intersect the line at infinity in 2 points, corresponding to the asymptotes—hyperbolas pass through infinity, with a twist. Going to infinity along one branch passes through the point at infinity corresponding to the asymptote, then re-emerges on the other branch at the other side but with the inside of the hyperbola (the direction of curvature) on the other side – left vs. right (corresponding to the non-orientability of the real projective plane)—and then passing through the other point at infinity returns to the first branch. Hyperbolas can thus be seen as ellipses that have been pulled through infinity and re-emerged on the other side, flipped.

Degenerate Cases

There are five degenerate cases: three in which the plane passes through apex of the cone, and three that arise when the cone itself degenerates to a cylinder (a doubled line can occur in both cases).

When the plane passes through the apex, the resulting conic is always degenerate, and is either: a point (when the angle between the plane and the axis of the cone is larger than tangential); a straight line (when the plane is tangential to the surface of the cone); or a pair of intersecting lines (when the angle is smaller than the tangential).

These correspond respectively to degeneration of an ellipse, parabola, and a hyperbola, which are characterized in the same way by angle. The straight line is more precisely a *double* line (a line with multiplicity 2) because the plane is tangent to the cone, and thus the intersection should be counted twice.

Where the cone is a cylinder, i.e. with the vertex at infinity, cylindric sections are obtained; this corresponds to the apex being at infinity. Cylindrical sections are ellipses (or circles), unless the plane is vertical (which corresponds to passing through the apex at infinity), in which case three degenerate cases occur:

Two parallel lines, known as a ribbon (corresponding to an ellipse with one axis infinite and the other axis real and non-zero, the distance between the lines), a double line (an ellipse with one infinite axis and one axis zero), and no intersection (an ellipse with one infinite axis and the other axis imaginary).

Eccentricity, Focus and Directrix

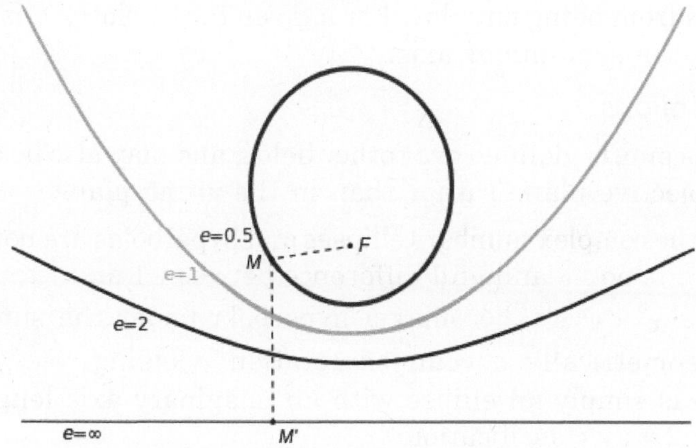

Figure: *Ellipse* (e=*1/2*)*, parabola* (e=*1*) *and hyperbola* (e=*2*) *with fixed focus* F *and directrix* $(e=\infty)$.

The four defining conditions above can be combined into one condition that depends on a fixed point *F*(the *focus*), a line *L*(the *directrix*) not containing *F*and a nonnegative real number *e*(the *eccentricity*). The corresponding conic section consists of the locus of all points whose distance to *F* equals *e* times their distance to *L*. For $0<e<1$ we obtain an ellipse, for $e=1$ a parabola, and for $e>1$ a hyperbola.

For an ellipse and a hyperbola, two focus-directrix combinations can be taken, each giving the same full ellipse or hyperbola. The distance from the centre to the directrix is a/e, where *a* is the semi-major axis of the ellipse, or the distance from the centre to the tops of the hyperbola. The distance from the centre to a focus is ae .

In the case of a circle, the eccentricity $e=0$, and one can imagine the directrix to be infinitely far removed from the centre. However, the statement that the circle consists of all points whose distance to F is e times the distance to L is not useful, because we get zero times infinity.

The eccentricity of a conic section is thus a measure of how far it deviates from being circular. For a given a, the closer a is to 1, the smaller is the semi-minor axis.

Generalizations

Conics may be defined over other fields, and may also be classified in the projective plane rather than in the affine plane.

Over the complex numbers ellipses and hyperbolas are not distinct, since there is no meaningful difference between 1 and –1; precisely, the ellipse $x^2 + y^2 = 1$ becomes a hyperbola under the substitution $y = iw$, geometrically a complex rotation, yielding $x^2 - w^2 = 1$ – a hyperbola is simply an ellipse with an imaginary axis length. Thus there is a 2-way classification:

Ellipse/hyperbola and parabola. Geometrically, this corresponds to intersecting the line at infinity in either 2 distinct points (corresponding to two asymptotes) or in 1 double point (corresponding to the axis of a parabola), and thus the real hyperbola is a more suggestive image for the complex ellipse/hyperbola, as it also has 2 (real) intersections with the line at infinity.

In projective space, over any division ring, but in particular over either the real or complex numbers, all non-degenerate conics are equivalent, and thus in projective geometry one simply speaks of "a conic" without specifying a type, as type is not meaningful.

Geometrically, the line at infinity is no longer special (distinguished), so while some conics intersect the line at infinity differently, this can be changed by a projective transformation – pulling an ellipse out to infinity or pushing a parabola off infinity to an ellipse or a hyperbola.

A generalization of a non degenerate conic in a projective plane is an oval. An oval is a point set that has the following properties, which are held by conics: 1) any line intersects an oval in none, one or two points, 2) at any point of the oval there exists a unique tangent line.

In other areas of Mathematics

The classification into elliptic, parabolic, and hyperbolic is pervasive in mathematics, and often divides a field into sharply distinct subfields. The classification mostly arises due to the presence of a quadratic form (in two variables this corresponds to the associated discriminant), but can also correspond to eccentricity.

Quadratic form classifications: quadratic forms Quadratic forms over the reals are classified by Sylvester's law of inertia, namely by their positive index, zero index, and negative index: a quadratic form in n variables can be converted to a diagonal form, as $x_1^2 + x_2^2 + \cdots + x_k^2 - x_{k+1}^2 - \cdots - x_{k+l}^2$, where the number of +1 coefficients, k, is the positive index, the number of –1 coefficients, l, is the negative index, and the remaining variables are the zero index m, so $k + l + m = n$. In two variables the non-zero quadratic forms are classified as:

- $x^2 + y^2$,– positive-definite (the negative is also included), corresponding to ellipses,
- x^2– degenerate, corresponding to parabolas, and
- $x^2 - y^2$– indefinite, corresponding to hyperbolas.

In two variables quadratic forms are classified by discriminant, analogously to conics, but in higher dimensions the more useful classification is as *definite,* (all positive or all negative), *degenerate,* (some zeros), or *indefinite* (mix of positive and negative but no zeros). This classification underlies many that follow.

Curvature: The Gaussian curvature of a surface describes the infinitesimal geometry, and may at each point be either positive – elliptic geometry, zero – Euclidean geometry (flat, parabola), or negative – hyperbolic geometry; infinitesimally, to second order the surface looks like the graph of $x^2 + y^2$,, x^2 (or 0), or $x^2 - y^2$.. Indeed, by the uniformization theorem every surface can be taken to be globally (at every point) positively curved, flat, or negatively curved. In higher dimensions the Riemann curvature tensor is a more complicated object, but manifolds with constant sectional curvature are interesting objects of study, and have strikingly different properties, as discussed at sectional curvature.

Second order PDEs: Partial differential equations (PDEs) of second order are classified at each point as elliptic, parabolic, or hyperbolic, accordingly as their second order terms correspond to an elliptic, parabolic, or hyperbolic quadratic form.

The behaviour and theory of these different types of PDEs are strikingly different – representative examples is that the Poisson equation is elliptic, the heat equation is parabolic, and the wave equation is hyperbolic.

Eccentricity classifications include: Möbius transformations Real Möbius transformations (elements of $PSL_2(R)$ or its 2-fold cover, $SL_2(R)$) are classified as elliptic, parabolic, or hyperbolic accordingly as their half-trace is $0 \leq | \mathrm{tr} | /2 < 1$, $| \mathrm{tr} | /2 = 1$, or $| \mathrm{tr} | /2 > 1$, mirroring the classification by eccentricity.

Variance-to-mean ratio: The variance-to-mean ratio classifies several important families of discrete probability distributions: the constant distribution as circular (eccentricity 0), binomial distributions as elliptical, Poisson distributions as parabolic, and negative binomial distributions as hyperbolic. This is elaborated at cumulants of some discrete probability distributions.

Cartesian Coordinates

In the Cartesian coordinate system, the graph of a quadratic equation in two variables is always a conic section – though it may be degenerate, and all conic sections arise in this way. The equation will be of the form

$$Ax^2 + Bxy + Cy^2 + Dx + Ey + F = 0 \text{ with } A, B, C \text{ not all zero.}$$

As scaling all six constants yields the same locus of zeros, one can consider conics as points in the five-dimensional projective space P^5.

Discriminant Classification

The conic sections described by this equation can be classified with the discriminant

$$B^2 - 4AC.$$

If the conic is non-degenerate, then:

- if $B^2 - 4AC < 0$, the equation represents an ellipse;
  - o if $A = C$ and $B = 0$, the equation represents a circle, which is a special case of an ellipse;
- if $B^2 - 4AC = 0$, the equation represents a parabola;
- if $B^2 - 4AC > 0$, the equation represents a hyperbola;
  - o if we also have $A + C = 0$, the equation represents a rectangular hyperbola.

To distinguish the degenerate cases from the non-degenerate cases, let " be the determinant of the 3×3 matrix [*A*, *B*/2, *D*/2 ; *B*/2, *C*, *E*/2 ; *D*/2, *E*/2, *F*]: that is, $\Delta = (AC - B^2/4)F + BED/4 - CD^2/4 - AE^2/$

4. Then the conic section is non-degenerate if and only if $\Delta \neq 0.$. If $\Delta=0$ we have a point ellipse, two parallel lines (possibly coinciding with each other) in the case of a parabola, or two intersecting lines in the case of a hyperbola.

Moreover, in the case of a non-degenerate ellipse (with $B^2 - 4AC < 0$ and $\Delta \neq 0.$), we have a real ellipse if $C\Delta < 0$ but an imaginary ellipse if $C\Delta > 0$. An example is $x^2 + y^2 + 10 = 0$, which has no real-valued solutions.

Note that A and B are polynomial coefficients, not the lengths of semi-major/minor axis as defined in some sources.

Matrix Notation

The above equation can be written in matrix notation as

$$[x \quad y].\begin{bmatrix} A & B/2 \\ B/2 & C \end{bmatrix}.\begin{bmatrix} x \\ y \end{bmatrix} + Dx + Ey + F = 0.$$

The type of conic section is solely determined by the determinant of middle matrix:

If it is positive, zero, or negative then the conic is an ellipse, parabola, or hyperbola respectively. If both the eigenvalues of the middle matrix are non-zero (i.e. it is an ellipse or a hyperbola), we can do a transformation of variables to obtain

$$\begin{pmatrix} x-a \\ y-c \end{pmatrix}^T \begin{pmatrix} A & \frac{B}{2} \\ \frac{B}{2} & C \end{pmatrix} \begin{pmatrix} x-a \\ y-c \end{pmatrix} = G$$

where a,c, and G satisfy $D + 2aA + Bc = 0, E + 2Cc + Ba = 0$, and $G = Aa^2 + Cc^2 + Bac - F$.

The quadratic can also be written as

$$[x \quad y \quad 1].\begin{bmatrix} A & B/2 & D/2 \\ B/2 & C & E/2 \\ D/2 & E/2 & F \end{bmatrix}.\begin{bmatrix} x \\ y \\ 1 \end{bmatrix} = 0.$$

If the determinant of this 3×3 matrix is non-zero, the conic section is not degenerate. If the determinant equals zero, the conic is a degenerate parabola (two parallel or coinciding lines), a degenerate ellipse (a point ellipse), or a degenerate hyperbola (two intersecting lines).

Note: that in the centred equation with constant term G, G equals minus one times the ratio of the 3×3 determinant to the 2×2 determinant.

As slice of Quadratic form

The equation

$$Ax^2 + Bxy + Cy^2 + Dx + Ey + F = 0$$

can be rearranged by taking the affine linear part to the other side, yielding

$$Ax^2 + Bxy + Cy^2 = -(Dx + Ey + F).$$

In this form, a conic section is realized exactly as the intersection of the graph of the quadratic form $z = Ax^2 + Bxy + Cy^2$ and the plane $z = -(Dx + Ey + F)$. Parabolas and hyperbolas can be realized by a horizontal plane ($D = E = 0$), while ellipses require that the plane be slanted. Degenerate conics correspond to degenerate intersections, such as taking slices such as $z = -1$ of a positive-definite form.

Eccentricity in Terms of Parametres of the Quadratic form

When the conic section is written algebraically as

$$Ax^2 + Bxy + Cy^2 + Dx + Ey + F = 0,$$

the eccentricity can be written as a function of the parametres of the quadratic equation. If $4AC = B^2$ the conic is a parabola and its eccentricity equals 1 (if it is non-degenerate). Otherwise, assuming the equation represents either a non-degenerate hyperbola or a non-degenerate, non-imaginary ellipse, the eccentricity is given by

$$e = \sqrt{\frac{2\sqrt{(A-C)^2 + B^2}}{\eta(A+C) + \sqrt{(A-C)^2 + B^2}}},$$

where $\eta = 1$ if the determinant of the 3×3 matrix is negative and $\eta = -1$ if that determinant is positive.

Standard form

Through change of coordinates these equations can be put in standard forms:

- Circle: $x^2 + y^2 = a^2$
- Ellipse: $x^2/a^2 + y^2/b^2 = 1$
- Parabola: $y^2 = 4ax$, $x^2 = 4ay$

- Hyperbola: $x^2/a^2 - y^2/b^2 = 1$, $x^2/b^2 - y^2/a^2 = -1$
- Rectangular hyperbola: $xy = c^2$

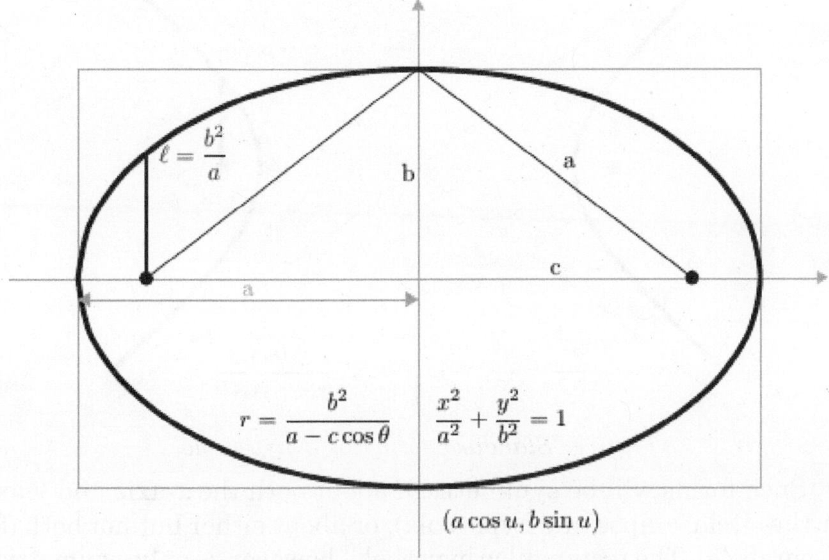

Figure: *Standard forms of an ellipse*

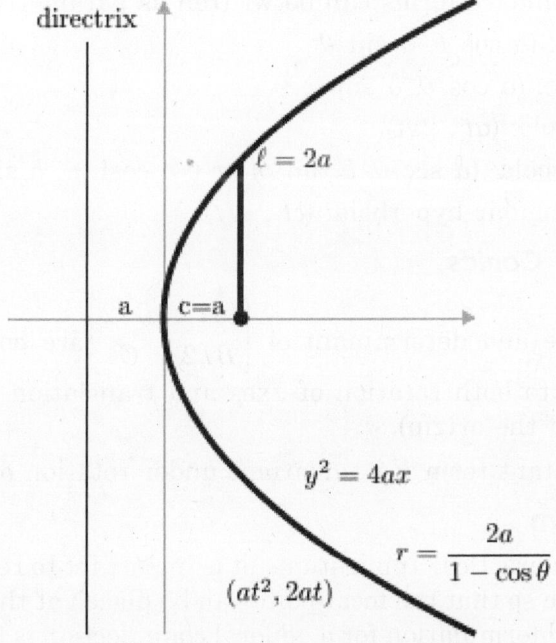

Figure: *Standard forms of a parabola*

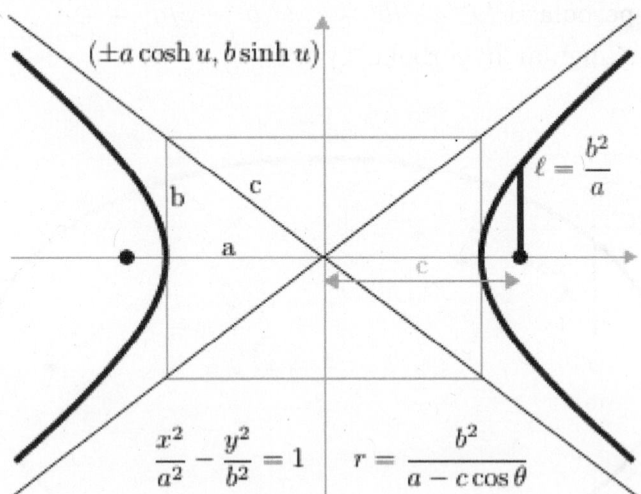

***Figure:** Standard forms of a hyperbola*

Such forms will be symmetrical about both the x-axis and y-axis (for the circle, ellipse and hyperbola), or about either but not both (for the parabola). The rectangular hyperbola, however, is only symmetrical about the lines $y = x$ and $y = -x$.

These standard forms can be written as parametric equations,

- Circle: $(a \cos \theta, a \sin \theta)$,
- Ellipse: $(a \cos \theta, b \sin \theta)$,
- Parabola: $(at^2, 2at)$,
- Hyperbola: $(a \sec \theta, b \tan \theta)$ or $(\pm a \cosh u, b \sinh u)$,
- Rectangular hyperbola: $(ct , c/t)$.

Invariants of Conics

The trace and determinant of $\begin{bmatrix} A & B/2 \\ B/2 & C \end{bmatrix}$ are both invariant with respect to both rotation of axes and translation of the plane (movement of the origin).

The constant term F is invariant under rotation only.

Modified form

For some practical applications, it is important to re-arrange the standard form so that the focal-point can be placed at the origin. The mathematical formulation for a general conic section is then given in the polar form by

$$r = \frac{l}{1 - e\cos\theta}$$

and in the Cartesian form by

$$\sqrt{x^2 + y^2} = (l + ex)$$

$$\Rightarrow \left(\frac{x - \frac{le}{1-e^2}}{\frac{l}{1-e^2}} \right)^2 + \frac{(1-e^2)y^2}{l^2} = 1$$

From the above equation, the linear eccentricity (*c*) is given by $c = \left(\frac{le}{1-e^2} \right)$.

From the general equations given above, different conic sections can be represented as shown below:

- Circle: $x^2 + y^2 = r^2$
- Ellipse: $\frac{\left(x - \sqrt{a^2 - b^2}\right)^2}{a^2} + \frac{y^2}{b^2} = 1$
- Parabola: $y^2 = 4a(x + a)$
- Hyperbola: $\frac{\left(x + \sqrt{a^2 + b^2}\right)^2}{a^2} - \frac{y^2}{b^2} = 1$

Homogeneous Coordinates

In homogeneous coordinates a conic section can be represented as:

$$A_1x^2 + A_2y^2 + A_3z^2 + 2B_1xy + 2B_2xz + 2B_3yz = 0.$$

Or in matrix notation

$$[x \quad y \quad z].\begin{bmatrix} A_1 & B_1 & B_2 \\ B_1 & A_2 & B_3 \\ B_2 & B_3 & A_3 \end{bmatrix}.\begin{bmatrix} x \\ y \\ z \end{bmatrix} = 0.$$

The matrix $M = \begin{bmatrix} A_1 & B_1 & B_2 \\ B_1 & A_2 & B_3 \\ B_2 & B_3 & A_3 \end{bmatrix}$ is called *the matrix of the conic section.*

$$\Delta = \det(M) = \det\left(\begin{bmatrix} A_1 & B_1 & B_2 \\ B_1 & A_2 & B_3 \\ B_2 & B_3 & A_3 \end{bmatrix}\right)$$ is called the determinant of the conic section. If $\Delta = 0$ then the *conic section* is said to be *degenerate*; this means that the conic section is either a union of two straight lines, a repeated line, a point or the empty set.

For example, the conic section

$$\begin{bmatrix} x & y & z \end{bmatrix}.\begin{bmatrix} 1 & 0 & 0 \\ 0 & -1 & 0 \\ 0 & 0 & 0 \end{bmatrix}.\begin{bmatrix} x \\ y \\ z \end{bmatrix} = 0$$

reduces to the union of two lines:

$$\{x^2 - y^2 = 0\} = \{(x+y)(x-y) = 0\} = \{x+y=0\} \cup \{x-y=0\}.$$

Similarly, a conic section sometimes reduces to a (single) repeated line:

$$\{x^2 + 2xy + y^2 = 0\} = \{(x+y)^2 = 0\} = \{x+y=0\} \cup \{x+y=0\} = \{x+y=0\}.$$

$$\delta = \det\left(\begin{bmatrix} A_1 & B_1 \\ B_1 & A_2 \end{bmatrix}\right)$$ is called the discriminant of the conic section.

If $\delta = 0$ then the *conic section* is a parabola, if $\delta < 0$, it is an hyperbola and if $\delta > 0$, it is an ellipse. A conic section is a circle if $\delta > 0$ and $A_1 = A_2$ and $B_1 = 0$, it is an rectangular hyperbola if $\delta < 0$ and $A_1 = -A_2$. It can be proven that in the complex projective plane CP^2 two conic sections have four points in common (if one accounts for multiplicity), so there are never more than 4 intersection points and there is always one *intersection point* (possibilities: four distinct intersection points, two singular intersection points and one double intersection points, two double intersection points, one singular intersection point and 1 with multiplicity 3, 1 intersection point with multiplicity 4). If there exists at least one intersection point with multiplicity > 1, then the two conic sections are said to be tangent. If there is only one intersection point, which has multiplicity 4, the two conic sections are said to be osculating.

Furthermore each straight line intersects each conic section twice. If the intersection point is double, the line is said to be tangent and it is called the tangent line. Because every straight line intersects a conic section twice, each conic section has two points at infinity (the intersection points with the line at infinity). If these points are real, the conic section must be a hyperbola, if they are imaginary conjugated, the conic section must be an ellipse, if the conic section has one double point at infinity it is a parabola. If the points at infinity are (1,i,0) and (1,-i,0), the conic section is a circle. If a conic section has one real and one imaginary point at infinity or it has two imaginary points that are not conjugated then it not a real conic section (its coefficients are complex).

Polar Coordinates

In polar coordinates, a conic section with one focus at the origin and, if any, the other on the x-axis, is given by the equation

$$r = \frac{l}{1 + e\cos\theta},$$

where e is the eccentricity and l is the semi-latus rectum. As above, for $e = 0$, we have a circle, for $0 < e < 1$ we obtain an ellipse, for $e = 1$ a parabola, and for $e > 1$ a hyperbola.

Pencil of Conics

A (non-degenerate) conic is completely determined by five points in general position (no three collinear) in a plane and the system of conics which pass through a fixed set of four points (again in a plane and no three collinear) is called a pencil of conics. The four common points are called the *base points* of the pencil. Through any point other than a base point, there passes a single conic of the pencil. This concept generalizes a pencil of circles.

In a projective plane defined over an algebraically closed field any two conics meet in four points (counted with multiplicity) and so, determine the pencil of conics based on these four points. Furthermore, the four base points determine three line pairs (degenerate conics through the base points, each line of the pair containing exactly two base points) and so each pencil of conics will contain at most three degenerate conics.

A pencil of conics can represented algebraically in the following way. Let C_1 and C_2 be two distinct conics in a projective plane defined over an algebraically closed field K. For every pair λ, μ of elements of K, not both zero, the expression:

$$\lambda C_1 + \mu C_2$$

represents a conic in the pencil determined by C_1 and C_2. This symbolic representation can be made concrete with a slight abuse of notation (using the same notation to denote the object as well as the equation defining the object.) Thinking of C_1, say, as a ternary quadratic form, then $C_1 = 0$ is the equation of the "conic C_1". Another concrete realization would be obtained by thinking of C_1 as the 3×3 symmetric matrix which represents it. If C_1 and C_2 have such concrete realizations then every member of the above pencil will as well. Since the setting uses homogeneous coordinates in a projective plane, two concrete representations (either equations or matrices) give the same conic if they differ by a non-zero multiplicative constant.

Intersecting two Conics

The solutions to a system of two second degree equations in two variables may be viewed as the coordinates of the points of intersection of two generic conic sections.

In particular two conics may possess none, two or four possibly coincident intersection points. An efficient method of locating these solutions exploits the homogeneous matrix representation of conic sections, i.e. a 3x3 symmetric matrix which depends on six parametres.

The procedure to locate the intersection points follows these steps, where the conics are represented by matrices:

- given the two conics $C_1$and C_2, consider the pencil of conics given by their linear combination $\lambda C_1 + \mu C_2$.
- identify the homogeneous parametres (λ, μ) which correspond to the degenerate conic of the pencil. This can be done by imposing the condition that $\det(\lambda C_1 + \mu C_2) = 0$ and solving for λ and μ. These turn out to be the solutions of a third degree equation.
- given the degenerate conic C_0, identify the two, possibly coincident, lines constituting it.
- intersect each identified line with either one of the two original conics; this step can be done efficiently using the dual conic representation of C_0
- the points of intersection will represent the solutions to the initial equation system.

Applications

Conic sections are important in astronomy: the orbits of two massive objects that interact according to Newton's law of universal gravitation are conic sections if their common centre of mass is considered to be at rest. If they are bound together, they will both trace out ellipses; if they are moving apart, they will both follow parabolas or hyperbolas. In projective geometry, the conic sections in the projective plane are equivalent to each other up to projective transformations.

Figure: *The paraboloid shape of Archeocyathids produces conic sections on rock faces*

Shaping Machine and Various Types of Shaping Machine

The shaper is a reciprocating type of machine tool used to produce flat surface. The surface may be horizontal, vertical or inclined. Modern shaper can generate contoured surfaces also. This is accomplished by holding the work rigidly on the machine and a reciprocating single point tool mounted on tool head.

The single point tool moves along a horizontal axis over the work piece. The metal working shaper was developed in the year 1836 by James Nesmyth an English man.

Purpose Purpose of the shaping machines are as follows:

- The main purpose of shaping machine is to machine plane horizontal surface.
- It is possible to form vertical and angular surfaces also. With proper tools and accessories even irregular surfaces can be formed.

Types of Shapers Shapers are classified as follows:

- *As Per Type of Motion.*
- Crank type
- Geared type
- Hydraulic type
- *As Per Position and Travel of Ram*
- Horizontal type
- Vertical type
- Travelling head type
- *As Per Design of Table*
- Standard shaper
- Universal shaper
- *As Per Cutting Stroke*
- Push type
- Draw type

Description of Various Shaping Machines

- *Crank Shaper:* The most common type of shaping machine is the crank shaper, in which the motion of its ram is being derived from a crank, a crank block, and a crank pin. The shaper size is known by the largest stroke of the ram. Usually it ranges from 12" to 36" with 24" being the most common.
- *Gear Type Shaper:* The reciprocating motion of the ram in this type of a shaper is obtained by means of a rack and pinion. The speed and the direction in which the machine will traverse depend upon the number of gears in the gear train. This type of shaper is not very widely used.
- *Hydraulic Shaper*: In a hydraulic shaper, the reciprocating movements of the ram are obtained by hydraulic power. Oil under high pressure is pumped into the operating cylinder fitted with a piston. The end of the piston rod is connected to the ram. The high pressure oil first acts on one side of the piston causing the piston to reciprocate and the motion is transmitted to the ram. The piston speed is changed by varying the amount of oil delivered by the pump. One of the most important advantages of this type of shaper is that the cutting speed and force of the ram drive are constant from very beginning to the end of the cut.

- *Horizontal Shaper:* In a horizontal shaper, the ram holding tool reciprocates in a horizontal axis. Horizontal shapers are mainly used to produce flat surfaces.
- *Vertical Shaper*: In a vertical shaper, the ram holding the tool reciprocates in vertical axis. In some of the vertical machines, provision is made to allow adjustment of the ram to an angle of about 10^0 from the vertical position. Vertical shapers may be crank driven, rack driven, screw driven, or hydraulic power driven.
- *Travelling Head Shaper*: In a travelling head shaper, the ram carrying the tool while it reciprocates, moves cross wise to give the required feed. Heavy jobs which are very difficult to hold on the table of a standard shaper and fed past the tool are held static on the basement of the machine while the ram reciprocates and supplies the feeding movements.
- *Standard or Plain Shaper:* A shaper is termed as standard or plain when the table has only two movements vertical and horizontal, to give the feed. The table may or may not be supported at the other end.
- *Universal Shaper:* In a Universal shaper, in addition to the two movements provided on the table of a standard shaper, the table can be swivelled about an axis parallel to the ram ways, and the upper portion of the table can be tilted to second horizontal axis perpendicular to the first axis. As the work is mounted on the table, it can be adjusted in different planes. A universal shaper is mostly used for tool room work.
- *Push Type Shaper*: This is the most general type of shaper used in common practice. The metal is removed when the ram moves away from the column, i.e. it pushes the work.
- *Draw Type Shaper*: In a draw type shaper, the metal is removed when the ram moves towards the column of the machine, i.e., draws the work towards the machine. The ram is generally supported by an overhead arm which ensures rigidity and eliminates deflection of the tool.

Shaping Machine Main Parts and Applications

Base: Base is a hollow casting upon which the other parts of the shaper are mounted. It also houses a reservoir for supply of oil which is circulated for moving parts of the machine.

Frame: Frame is a large casting that rests upon the base. It houses the mechanism and supports the ram and the table.

Ram: Ram is the main moving part of the shaper. It holds and drives the single point cutting tool back and forth across the work. This is attached to the rocker arm through a small link lever, and slides along the top ways.

Tool Head**:** The tool head is designed to hold the cutting tools. It is attached to the front of the ram. It may be swivelled to a required angle to the left or right with respect to vertical slide of tool head and locked in place with the help of a bolt or screw. It may also be adjusted vertically and locked in position. The tool post is fastened to a clapper box to ride over the work when the ram returns back. It is lifted upward to relieve the friction, which will prevent the tool cutting edge from dragging on the work-piece. This also prevents tool wear and poor finish.

Table: It is a metal box, which is attached to the frame of the shaper. It has "T" slots on the top and sides, which are used for clamping work or a vice on the table. It slides on the front side through a saddle. It is supported in the front by a bracket attached to the base which is called the out-board clamp.

The table is often connected with a link lever and a ratchet from the rocker arm, which imparts the automatic movement when required. This automatic movement may be of definite amount during each return stroke of the ram.

Though the table can be vertically fed to give a cut on some special occasions, but this method of feeding is not generally used since it is too laborious. Table is vertically moved only to accommodate jobs that vary considerably in heights. This type of table cannot be swivelled or turned.

Universal Table: A universal table is the one that can be operated in the same manner as a standard table. In addition, it may be swivelled to the left or right and if needed can be rotated to a full circle. A dial plate graduated in degree indicates the angular setting. The top surface may be tilted up to 15° forward or backward.

Table Support or Out Board Clamp**:** It gives an extra support to the table. It is fitted on the front side of the machine to take up the cutting pressure as the tool cuts in forward motion. To lower or raise the table the clamping nuts of the out board support is to be loosened.

Pointer or Stroke Indicator: It indicates the length of stroke. When tightened, it clamps the ram where by the oscillating motion of the rocker arm is transferred to the ram through the link lever. If it is not clamped, movement of rocker arm will not be transferred to the ram.

Friction Quadrant or Automatic Movement of Vertical Tool: This mechanism is incorporated in heavy machines to impart automatic feed of the tool through the swivelling slide.

***Small Link*:** This link is connected to one end of the rocker arm and the other end to the ram. The oscillating movement of rocker arm through the link is converted into the reciprocating motion of the ram. It is pivoted on the sides to give freedom of movement when the rocker arm moves forward and backward.

***Rocker Arm or Long Link Lever*:** The rocker arm is pivoted at the bottom of the machine, which carries the sliding block in its slot. The rotary motion of the bull gear is converted into oscillating motion of the rocker arm through the crank block, crank pin, and sliding block.

***Bull Gear*:** It is a big (generally helical) gear which carries the mechanism and is driven by a pinion. On the face of this bull gear, there is a dovetail way in which the crank block is held.

***Sliding Block*:** The block is made of wear resisting metal. It is generally made out of bronze and slides in the slot of the rocker arm with up and down movement.

***Elevating Screw*:** It is used for raising and lowering the table.

***Stroke Adjustment Crank*:** Through this the stroke adjustment is done by off setting the crank block from the centre of the bull gear to the required distance.

***Clapper Box*:** It is hinged on the swivelling slide and carries the tool holder.

Keys, Cotters and Joints Used in Advanced Engineering Drawing

Introduction to Keys, Cotters and Joints

A machine runs by the power supplied to it by a prime mover such as, motor, engine, etc. The power is transmitted from the prime mover to the machine through a coupler which couples the shafts of the prime mover and the machine. The most commonly employed method to connect a shaft and a part is to drive a small piece of metal, known as *key* between the shaft and the hole made in the part mounted over it. The key will be driven such that it sits partly into the shaft and partly into the part mounted on it. To introduce the key, axial grooves, called *key ways* are cut both in the shaft and the part mounted on it. The key is fitted between the shaft and the part mounted over it. While transmitting the power, the key will be subjected to shear and crushing forces. Keys are extensively used to hold pulleys, gears, couplings, clutches, sprockets, etc., and the shafts rigidly so that they rotate together. They are also used to mount the milling cutters, grinding wheels, etc., on their spindles.

Rods of circular and square or rectangular cross sections subjected to axial tensile or compressive forces, are connected together by a *cotter joint*. It is a temporary method of fastening of the two rods which will have to be frequently assembled and disassembled. Its chief advantages are that the joint can be quickly assembled and disassembled and the rods occupy exactly the same relative positions after assembly. The cotter joints are used to connect the piston rod to the cross head of the steam engine, pump or compressor. Long tie bars in steel structures are sometimes built up of round bar, of short lengths and joined together by cotter joints.

Keys

A *key* is a square, rectangular, circular or semi-circular piece of mild steel or wrought iron, which is inserted in a recessed shaft or hub called *key ways*. When the key is in position, it prevents relative rotary motion between the mating parts. Sometimes, a key also prevents axial motion in the two parts.

Types of Keys

The keys can be classified as per the shape and purpose for which they are used. Following are the types of keys.

(a) Sunk keys

(b) Saddle keys

(c) Tangent keys

(d) Serrated shaft and Splines

(e) Round keys or Taper pins

Classification of Keys

Keys are classified into three types as follows:

(a) Taper keys

(b) Parallel or Feather keys

(c) Special keys

Taper Keys: A *taper Key* is of rectangular cross section having uniform width and tapering thickness. The taper keys are used to transmit only the turning moment between the shaft and the hub without any relative rotational and axial motion between them. The examples of tapered keys are

(a) Taper sunk key

(b) Saddle key

(c) Flat key

(d) Gib head key

Parallel or Feather Keys: A *parallel key* or *feather key* is also of rectangular cross section of uniform width and thickness throughout. Parallel keys are used to transmit the turning moment between the shaft and the hub along with the provision to allow a small sliding axial motion between them wherever required. The examples of the parallel keys are:

(a) Parallel sunk key

(b) Peg key

(c) Single head key

(d) Double head key

(e) Spline shaft.

Special Keys: The *woodruff key, cone key* and *pin key* are the special purpose keys used for specific applications.

Sunk Keys: A key that engages a slot formed in both pulley and shaft is known as *sunk keys.* The sunk keys are of following types:

(a) Rectangular key

(b) Square key

(c) Gib head key

(d) Woodruff key

(e) Feather key

Rectangular Key: This type of key is rectangular in cross section and is very commonly used. The dimensions of this type of key are as follows:

Width of the key $W = D / 4$

Thickness of the key $T = D / 6$

Where D is the diametre of the shaft.

A taper of 1 in 100 on the thickness and parallel in width is given to the key. The taper is given on the upper surface of the key, i.e. the hub side. The recess in the hub is also given the same taper. The thickness of the key is measured at the large end.

Square Key: The cross section of the key is square and the dimension is:

Side of the key = *D / 4*

A taper of 1 in 100 on the thickness and parallel in width is given to the key. The taper is given on the upper surface of the key, i.e. the hub side. The recess in the hub is also given the same taper. This type of key is used for mounting pulleys or gears on shafts. Square keys are used for shafts upto *17mm* in diametre.

Gib Head Key: This key can be rectangular or square in cross section having a head at the large end. The head makes it easier to remove the key from the hub and shaft. The slot for gib head key must have an open end to permit assembly. For this reason it is placed at the end of a shaft. The dimensions of a gib head key.

Dia of the Shaft = D

Width of Key 'W' = D/4

Thickness of the key 'T' =D/6

Taper on thickness = 1 in 100

h = 1.75 T, b = 1.5 T

Woodruff Key: A woodruff key is segmental in shape and is an easily adjustable sunk key. The key fits into a semicircular key way cut into the shaft. The top of the key fits into a plain rectangular keyway in the hub.

This key has the advantage of aligning itself with the taper of the hub and will not easily turn over, because of its extra depth in the shaft. Woodruff key largely used in automobile work and machine tools. The dimensions of the woodruff key are standardized. The dimensions of a woodruff key for *ψd* and *ψ22* shaft. A woodruff key is designated as:

Woodruff key 5 X 9 IS: 2294

Where 5mm is the width and 9mm is the height of the key. IS: 2294 is BIS code for woodruff key.

Feather Key: The feather key is attached to the shaft or the hub and permits relative axial movement.

Saddle Keys: They are tapered keys and are of two types :

(a) Flat saddle key

(b) Hollow saddle key

Flat Saddle Key: It is a tapered key which fits in the key way of the hub and the flat surface on the shaft. It has got the tendency to slip round the shaft, that is why, it is suitable for light duty.

d = Dia of the Shaft, T = d/12, W = d/4

Hollow Saddle Key: This is also a tapered key fitting into the key way of the hub and the bottom of the key is curved so as to fit the curved surface of the shaft. It is suitable for light duty work.

d = Dia of the Shaft, T = d/12, W = d/4

Tangent Keys

These types of keys are used for heavy duty. The key is placed tangential to the shaft. These keys can withstand torsion on one side. If it is required to be used in a reversible unit then two keys at 90^0 or 120^0 are used. The key can have two wedge shaped pieces or a single rectangular or square piece.

Serrated Shafts and Splines

To prevent the key working loose, to reduce the working stresses and to give a greater bearing surface, it is now a standard practice to design shafts with keys machined integrally with the main body. The *splines* or *serrations* are milled or hobbed to size and shape and then heat treated. The slots or serrations are finished by broaching. The serrated or splined shafts are generally used in automobiles. In case of splined shaft, the number of splines can be four, six or even more.

Round Key or Taper Pin

A round or taper pin is commonly used for fastening a collar and pulley to shaft. The hole for the pin is drilled and reamed with the parts assembled. In some instances the pin is expected to shear before other parts of the assembly are damaged. So it also acts as a safety device.

Dia of Shaft= D Dia of Taper Pin d = D/6

Cotters

A cotter is a flat, wedge shaped piece of steel, used to connect rigidly, two parts which are subjected to axial forces only. The cotters have a uniform thickness throughout the length and taper on one or both side. The taper varies from 1 in 8 to 1 in 48. The cotters having large tapers need a locking device.

Length of the cotter L = *3.5d* to *4d*

Breadth of the cotter B = *d* to *1.32d*

Thickness of the cotter T = *0.15d* to *0.25d*

The cotter joints are subjected to axial forces such as tensile or compressive forces. The joints differ from key joints which are used to join shafts subjected to torsional stresses only.

Comparison between Keys and Cotters

The main difference between keys and cotters are as follows:

(a) Keys are driven parallel to the axis whereas cotters are driven perpendicular to the axis.

(b) Keys are used in parts subjected to torque whereas cotters are used in parts subjected to tensile or compressive force.

(c) Keys resist shear over a longitudinal section whereas cotters resist shear over two transverse sections.

Cotter Joints

Cotter joints are used to connect together two rods rigidly to transmit power in the length-wise direction. Sometimes when it is required to increase the length of the rod or to connect a rod directly to the machine to transmit power through the rod as in the case of a connecting rod end of a steam engine, a joint is used. Cotter joints can be used for round or square rods. Some of the joints are described below.

Sleeve and Cotter Joint: This type of joint is used for light transmission of axial loads from one rod to another rod. The ends of the two rods are forced into a sleeve. Two slots are provided in the sleeve, and one slot each is provided in the rods to take one cotter each.

The slot considerably decreases the strength of the rod, that is why the diametre of the rods is increased for the length which comes inside the sleeve.

Spigot and Socket Cotter Joint: This type of joint is used for round rods. The end of one rod is forged in the shape of a socket and the other in the shape of a spigot. Slots are provided in the spigot and socket ends to accommodate the cotter. The diametre of the spigot end is increased to overcome the loss of strength due to the slot. To make the joint rigid and perfectly tight, the slots are made slightly out of alignment, so that when the cotters is driven in, it will tend to force the spigot and socket ends towards each other.

Gib and Cotter Joint for Square Rods: This type of joint is used for joining two square rods. One end of the rod is forged in the shape of a fork. The other rod is pushed into the fork. Slots are provided in the fork and the rod to accommodate the *gib* and *cotter* while assembling the parts. The gib is inserted first so that the straight surface touches the slot of the fork and then cotter is hammered into the rest of the slot. Care should be taken that the tapered side of the gib and cotter should be face to face with each other.

Screw Threads and Threaded Fasteners

A *screw* is a cylindrical or a conical rod with a helical groove cut on it. Its function is to transform the input rotary motion into output reciprocating motion. A *screw thread* is a continuous helical ridge formed by cutting a helical groove on a cylindrical or conical shank.

Basically screw threads are used to fasten the parts together or to transmit motion and so power, or for the relative adjustments of the adjacent parts. To perform this specific function, threads of different profiles are used. The shape of the different types of the standard thread profiles with proportions and their suitability for specific purpose are discussed here.

Screw Thread Terminology

The various elements of a screw thread.

Conventional Representation of Screw Threads

Since the true representation of screw threads involves drawing of the actual helical curves which requires laborious drafting work and time, threads are shown conventionally in the simplified form.

Representation of V-Threads

The external V- threads are represented by two thin continuous lines on the non circular view, and by a thin incomplete circle for about three-fourths of its circumference on the circular view. If accurate representation is required, the thin lines must be drawn at a distance equal to the actual root diametre. For approximate representation, they may be drawn at a distance equal to O.9x diametre of the rod. The limit length of the threads is shown by a thick line or by a medium dashed line.

When the externally threaded parts are to be shown in section, the thin lines and the incomplete thin circle representing the root diametre must be drawn on the hatched area. The internal V- threads are represented.

Representation of Square and Other Forms of Threads

The square and other forms of threads may be represented in the same way as the V- threads. However, the form of the thread may be indicated.

Designation of Thread

Threads are designated by indicating the type of thread, the major diametre and the pitch. V-threads of ISO profile are designated by the letter M followed by the major diametre and the pitch, the two being separated by x symbol.

For example, V-threads of major diametre 10 mm and pitch 1.25 mm is designated as M10 x1.25. Sometimes when the V-threads are designated without indicating the pitch, for example M 10, it will mean that the threads are cut with coarse pitch, whose value is to be obtained from the relevant Indian Standard Codes.

Multi start threads are designated by specifying the number of starts. The screw threads are always considered as right hand unless otherwise specified. When left hand threads are to be designated the abbreviation LH must be used.

The square threads are designated by the ¡% symbol. The trapezoidal threads are designated by the letters Tr.

Bolts and Nuts

Chamfering on Bolt Heads and Nuts: The sharp corners on the external flat end faces of bolt heads and nuts are taken off by chamfering conically at 30° to the flat faces.

The chamfering produces hyperbolic curves on the side faces and leaves circular end face. The bolt head is chamfered on its flat end face. The nut is chamfered either only on one end, or on both of its ends. The diametre of the chamfer circle will be equal to the distance across flats of the hexagon.

Countersinking of Nuts

To facilitate easy insertion of the nut over the bolt at its threaded end, the threaded holes in the nut are countersunk either only at the bearing face of the nut or on both the faces.

Drawing of Hexagonal Bolt and Nut

In machine drawing practice, hexagonal bolts and nuts are drawn almost in all the drawings. Drawing of these to the actual dimensions involves laborious drafting work and unnecessary time.

Since the bolts and nuts are standard items, they need not be drawn in the assembly drawings to accurate sizes, instead they are drawn conventionally to empirical proportions listed in table. All the proportions are expressed in terms of the major diametre of the bolt and nut.

Table: *Empirical Proportion of Hexagon and Square Head Bolt and Nut*

Detail	*Proportion*
Nominal Diameter	d = Size of Bolt or Nut, mm
Width Across Flats	s = 1.5d + 3mm
Width Across Corners	c = 2d
Thickness of Bolt Head	k = 0.8d
Thickness of Nut	m = 0.9d
Root Diameter	d_1 = d — (2 x Depth of Thread) or = d — (4 x Thickness of lines) or = 0.9d (approximate)
Length of Bolt	l = As specified
Thread Length	b = 2d + 6mm (for l < 150mm) = 2d + 12mm (for l > 150mm)
Radius of Bolt End	r = d (for spherical ends)
Chamfer of Bolt End	z = Depth of Thread x 45° or = 0.1d (Approximate)
Chamfer Angle of Bolt Head & Nut	= 30°

Step by Step Procedure for Drawing of Hexagonal Bolt and Nut

Step 1. Draw the shank of the bolt equal to the given diametre *d* and length. The thickness of bolt head equal to *0.8d* and the thickness of nut equal to *0.9d* are marked. Measure the width across corners equal to *2d* and complete the three faces of the bolt head and the nut in thin lines. The right view of the bolt and nut assembly is drawn as follows. With any point C1 on the axis as centre and radius equal to d, draw a thin circle. Draw the vertical diametre 1- 2 of this circle. With 1 and 2 as centres and radius equal to d cut the circle on either side of the vertical axis and inscribe the hexagon. The chamfer circle is drawn as a thick circle with the centre C1 and radius C1E.

Step 2. The *chamfer arcs* in the three face view of bolt head and nut are drawn as follows. Through the corner B, draw a line at 30° to the axis of the bolt or nut to cut it at O1. With O1 as centre and radius O1A draw the chamfer arc in the centre face. The chamfer arcs on the two side faces are drawn as follows. Draw the perpendicular bisector of BC to cut BO1 at O2. With O2 as centre and radius O2D draw the chamfer arc. Repeat the construction on the other side face.

Step 3. The *chamfer lines* on the side faces of the *three face views* of the bolt head and nut are drawn as follows. Through the points P and Q draw lines inclined at 30° to the flat face of the bolt head or nut. The end of the bolt is chamfered to *0.1 d x 45°*. The threaded portion of the shank is indicated by drawing two thin lines at a distance equal to *d1= 0.9d*. The *root circle* in the right view is represented by a thin three-fourth of a circle drawn with centre C1 and diametre 0.9d. The *two face view* of the bolt head and the nut is projected from the *side view*. If the side view is not drawn, then the distance across the flats is measured equal to *1.5d + 3mm*. The chamfer arcs in the two face view are drawn as follows. Project P to get X. Mark the midpoint of FG as Y. Draw the perpendicular bisector of XY and FG to intersect each other at O3. With centre O3 and radius O3Y, draw the chamfer arc. Repeat the construction on the other face.

Drawing of Square Head Bolt and Nut

The square head bolts and nuts are drawn to the proportion shown in table.

Step By Step Procedure for Drawing Square Bolt and Nut

Step 1. Draw the *shank* of the bolt equal to the given *diametre d* and the length of the bolt. The *thickness* of the bolt head equal to *0.8d* and the thickness of the nut equal to *0.9d* are marked. The right view of the bolt and the nut assembly is drawn as follows. With any pointC1 on the axis as centre and diametre equal to *1.5d + 3mm* draw the *chamfer circle*. Draw a square circumscribing the chamfer circle with its sides inclined at 45° to the axis. Project the corners 1 and 2 to get points P. Draw a *thick circle* with *diametre d* to indicate the nominal diametre.

Step 2. The *chamfer arcs* in the view across the corners of the bolt and nut are drawn as follows. Through the corner P, draw a line inclined at 30° to the axis. Draw the perpendicular bisector of PQ to intersect the 30° line at O. With O as centre and radius OR draw the chamfer arc. Repeat the construction on the other face.

Step 3. The *chamfer line* is drawn at 30° to the flat face of the bolt head and nut. The threaded portion on the shank of the bolt is indicated conventionally by drawing two thin lines spaced at a distance equal to the *root diametre d1=0.9d.* The *root circle* in the right view is represented by a thin three-fourth of a circle drawn with centre C1 and diametre *0.9d.* The end of the bolt is chamfered to *0.1 d x 45°.*

Rivets and Riveted Joints

Rivets are used as permanent fasteners. Riveted joints are extensively used for bridges, roof trusses, boilers, tanks, ship building, etc. These joints have strength, rigidity and can be made leak- proof. Welding has become very popular, but riveting still enjoys its application in many fields, especially where welding is difficult or impracticable.

Rivets

A rivet is a cylindrical solid or hollow bar with a *head* on one side. The middle portion is known as *body* or *shank.* The bottom part is called *tail.*

Forms and Proportions of Rivet Heads

Various types of rivet heads for use in general engineering work and boiler work as recommend-ed by the Bureau of Indian Standards. (For reference only. Not for examination purpose). The different proportions of these rivet heads are given in terms of the nominal diametre *'d'* of the rivet.

The rivet heads to be used for general purpose for diametres below 12 mm are specified in the Indian Standard Code *IS: 2155-1962* and for diametres between 12 and 48 mm are specified in the Indian Standard Code *IS: 1929-1961.*

The rivet heads to be used for boiler work are specified in the Indian Standard Code *IS: 1928-1961.* The rivet heads to be used for ship building are specified in the Indian Standard Code *IS: 4732-1968.*

Riveted Joints

The two types of riveted joints are:

Lap Joints: The simplest way to connect two plates by riveting is to overlap them over a short distance along their edges and drill a row of holes through both of them in the overlapped portion and a joint is made by riveting. This type of joint is called a *lap joint.*

Butt Joints: When the plates are intended to take heavy loads, the buckling action of the plates may be avoided by placing the connecting plates in alignment so as to butt each other. A *cover plate* called *strap* is placed over the joint between them and riveted to hold both the plates together. This type of joint is called *butt joint.*

Different Types of Lap Joints

When only one row of rivets is used to make a lap joint, it is called *single riveted lap joint.* When the lapping width of the plates is more, one row of rivets may not be sufficient to withstand the shearing load and prevent the buckling or the lapping plates. In such cases, to prevent the buckling of the lapping plates one more row of rivets are used. Since two rows of rivets are used to make the lap joint; it is called *double riveted lap joint.* Similarly if three rows of rivets are used to make a joint, it is called *treble riveted lap joint.* When more than one row or rivets are used as in double and treble riveted lap joints, the rivets in the adjoining rows may be arranged either in *chain* or *zig-zag* formation. In chain riveting, the rivets in the adjoining rows are placed directly opposite to each other i.e., the centre of the rivets in the adjoining rows lie in a line perpendicular to the row of rivets. In zig-zag riveting, the rivets in the adjoining rows are staggered to place the rivets in between those of the previous row. The chain riveting is generally used in the structural work, while the zig-zag riveting is used in the boiler construction. The chief advantage of the chain riveting is that a greater number of rivets can be provided per metre length than is possible with zig-zag riveting, and with regard to the joints connecting thin plates this may be an important factor in order to reduce the bearing pressure on the rivets.

Different Types of Butt Joints

Although two rows of rivets are used to make a simple butt joint, since each of the butt plates are connected to the cover plates by only one row of rivets; the joint is called *single riveted butt joint with double cover plates.* If two rows of rivets are used for each of the butt plates, the joint is called *double riveted butt joint with double cover plates.* Similarly if three rows of rivets are used for each of the butt plates, the joint is called *treble riveted butt joint with double cover plates.* The rows of rivets in double and treble riveted butt joints may be arranged either in *chain* or *zig-zag* form.

Welding Joints

Like riveting, welding is also a method of making permanent joints between two metal plates. Welding has become a common practice in the fabrication of cylindrical forms such as pressure vessels and boilers. Standard steel shapes, bars and plates are welded together to make frames and bases of machine tool, jigs and fixtures and other steel structures.

To make complicated machine parts welding is preferred over casting and forging. The process of welding is simpler and cheaper as it eliminates the cost of patterns and dies.

The automobile, aircraft and ship building industries have developed welding as a major fabricating method because practically all ferrous and non ferrous metals can be welded with ease.

Welding is also used as a repair medium, e.g. to reunite metal at cracks, to build up a small part that has broken of such as gear tooth or to repair a worn out surface of bearing.

Types of Welding Joints

There are five basic types of welded joints.

(a) Lap joint

(b) Butt joint

(c) T – joint

(d) Corner joint

(e) Edge joint

Welding Symbols

All drawings having welded joints are marked with certain symbols known as *weld symbols*. The following symbols have been prescribed by the Bureau of Indian Standards.

(a) Elementary welding symbols

(b) Supplementary welding symbols

(c) Combination of elementary symbols

Elementary Welding Symbols: The various types of welds are marked on the drawing by a symbol which is normally similar to the shape of the weld. The symbol does not mention the process to be employed for the weld.

Table: *Elementary Welding Symbols*

Sl No	DESIGNATION	ILLUSTRATION	SYMBOL
1	BUTT WELD BETWQEEN PLATES WITH RAISED EDGES; EDGE FLANGED WELD/USA/ (THE EDGES BEING MELTED DOWN COMPLETELY)		ᴖᴖ
2	SQUARE BUTT WELD		||
3	SINGLE BUTT WELD		V
4	SINGLE-BEVEL BUTT WELD		V
5	SINGLE-V BUTT WELD WITH BROAD ROOT FACE		Y
6	SINGLE-BEVEL BUTT WELD WITH BROAD ROOT FACE		Y
7	SINGLE BUTT WELD (PARALLEL OR SLOPING SIDES)		Y
8	SINGLE-J BUTT WELD		μ
9	BACKING RUN; BACK OR BACKING WELD / USA/		◡
10	FILLET		◺
11	PLUG WELD; PLUG OR SLOT WED/USA/		⊓
12	SPOT WELD	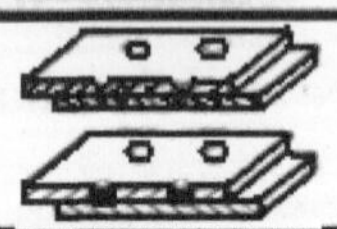	O
13	SEAM WELD	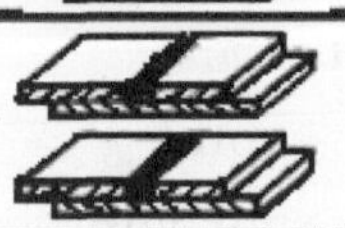	⊖

Supplementary Welding Symbols: When the shape of external surface of the weld is also important then it is mentioned along with elementary symbols.

Table: *Supplementary Symbols*

Shape of Weld Surface	***Symbol***
FLAT (FINISHED FLUSH)	%
CONVEX	“#
CONCAVE	##

Combination of Elementary Symbols: When two types of welding shapes are combined in one joint then the prescribed elementary symbols are also combined together and mentioned on the drawing.

***Table:** Combination of Elementary Symbols*

DESIGNATION	ILLUSTRATION	SYMBOL
FLAT SINGLE-V BUTT WELD		
CONVEX DOUBLE V-BUTT WELD		
FLAT SINGLE V-BUTT WELD WITH FLAT RUN		
CONCAVE FILLET WELD		

Essentials of a Welding Symbol

As per the Bureau of Indian Standards the weld symbols on each joint in the drawing must have the following:

(a) Arrow Line

(b) Reference Line

(c) Welding Symbols

Arrow Line: An inclined line with an arrow head is drawn pointing towards the joint. As all weld joints have two sides, the sides having the arrow is known as "arrow side" and the other side of the joint is known as "other side".

In the case of cruciform joint with fillet welds, the joint is divided into two parts 'A' and 'B' and the weld is marked as "arrow side "of joint 'A' or "other side" of joint 'A'. Similarly for joint 'B'.

The position of the arrow line with respect to the weld joint is not of much importance. In case of joints 4, 6 and 8 the arrow line must point towards the prepared plates.

Reference Line: The arrow line is joined by line known as *reference line* drawn parallel to the bottom edge of the drawing. If it is not possible to draw a parallel line then it is drawn perpendicular to the bottom edge of the drawing. The reference line consists of two lines, one continuous and the other dashed (dotted). The dashed line can be drawn above or below the continuous line. The dashed line represents the 'other side'. In case of symmetrical welds on both sides of the joint, the dotted line is omitted.

Welding Symbols: The proper weld symbols according to the type of the weld is chosen from tables. It is placed above or below the reference line when the weld is on 'arrow side' of the joint. The symbol placed on the dashed line represents weld is on the 'other side' of the joint. In the case of a symmetrical weld the symbol is placed both above and below the reference lines.

Dimensions

Each weld symbol is accompanied by dimensions of the cross-section and the length of weld. The dimensions of the cross section of the weld are placed on the left hand side of the weld symbol and the length of the weld is placed on the right hand side of the symbol. However, if the whole length of the joint is to be welded then the length is omitted. As the cross section of each type of weld is different, the dimensioning method changes accordingly. Some types of weld are dimensioned as shown in table.

Screw Threads and Threaded Fasteners

A *screw* is a cylindrical or a conical rod with a helical groove cut on it. Its function is to transform the input rotary motion into output reciprocating motion. A *screw thread* is a continuous helical ridge formed by cutting a helical groove on a cylindrical or conical shank. Basically screw threads are used to fasten the parts together or to transmit motion and so power, or for the relative adjustments of the adjacent parts. To perform this specific function, threads of different profiles are used. The shape of the different types of the standard thread profiles with proportions and their suitability for specific purpose are discussed here.

Screw Thread Terminology

The various elements of a screw thread. Screw threads, nuts and bolts have been discussed.

Conventional Representation of Screw Threads

Since the true representation of screw threads involves drawing of the actual helical curves which requires laborious drafting work and time, threads are shown conventionally in the simplified form.

Representation of V-Threads

The external V- threads are represented by two thin continuous lines on the non circular view, and by a thin incomplete circle for about three-fourths of its circumference on the circular view. If accurate representation is required, the thin lines must be drawn at a distance equal to the actual root diametre. For approximate representation, they may be drawn at a distance equal to O.9x diametre of the rod. The limit length of the threads is shown by a thick line or by a medium dashed line.

When the externally threaded parts are to be shown in section, the thin lines and the incomplete thin circle representing the root diametre must be drawn on the hatched area. The internal V- threads are represented.

Geometric Constructions Project

Geometrical constructions are very essential in the preparation of engineering drawing. Every draftsman should be familiar with them and must be able to apply them. Geometrical constructions are mostly based on plane geometry. There are numerous geometrical constructions. Some basic constructions which are used frequently in engineering drawing are explained in this chapter. Since the subject of plane geometry is a pre-requisite for a course in the engineering drawing, the mathematical proofs are omitted here.

Straight Lines: A straight line is the shortest distance between two points.

Example

Bisect a given straight line.

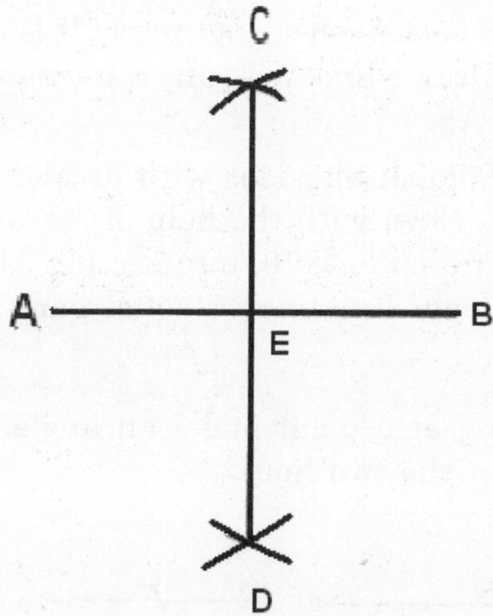

Figure: *Bisecting Straight Line*

Solution: Let AB be a given straight line. It is to be bisected. Taking centre A and radius greater than half AB, draw arcs on both the sides of AB. Again, taking centre B and the same radius, draw arcs on both the sides of AB, intersecting the previous arcs at C and D. Draw a line joining C and D intersecting AB at E.

Then CD bisects the line AB at E.

Thus, AE = EB = 1/2 AB

Example

Divide a given straight line into any number of equal parts.

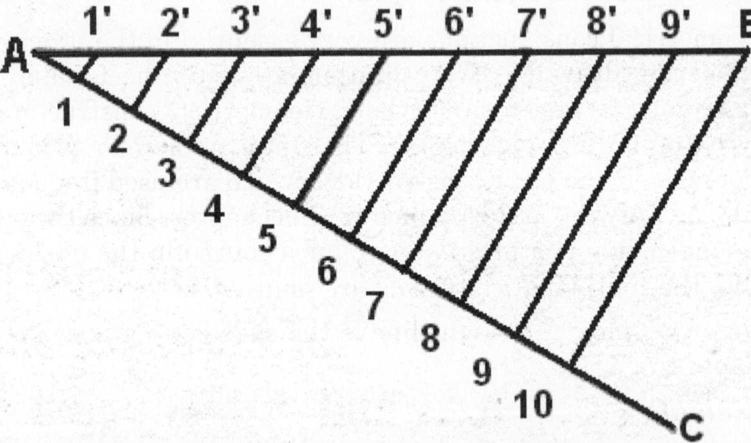

Figure: *Dividing a Straight Line into Equal Parts*

Solution: Let AB be a given straight line. It is to be divided into say, ten equal parts. Draw a line AC of any convenient length inclined at an acute angle to AB.

From A, mark 10 equal divisions with divider on AC as 1, 2, 3 ... 10. Join B and 10. Now, with the help of setsquares draw lines through 1, 2, 3 etc. parallel to B-10, intersecting AB at 1', 2', 3', etc. thus dividing the straight line into 10 equal parts.

Angle

Two lines meeting at a point make an angle. An angle is the included inclination of the two lines.

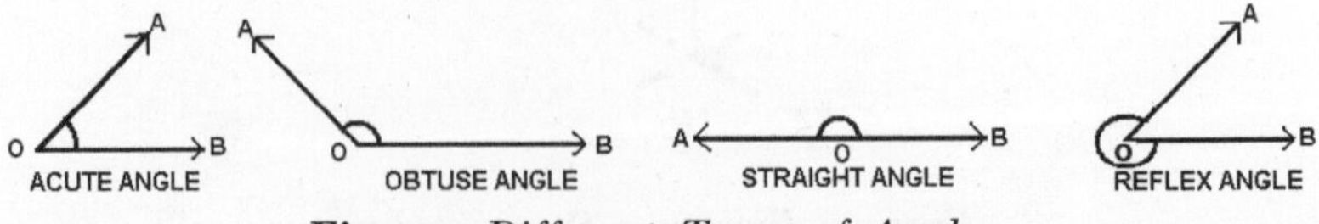

Figure: *Different Types of Angles*

Example

Bisect a given angle.

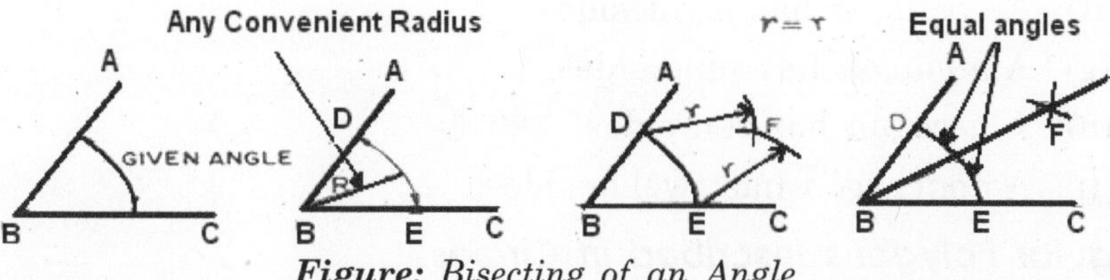

Figure: *Bisecting of an Angle*

Solution: Let ABC be a given angle. It is to be bisected. Taking centre B and any convenient radius, draw an arc cutting AB and BC at D and E respectively. Now, taking centre D and E and any convenient radius, draw arcs intersecting each other at F. Join B and F. Then BF is the bisector of the angle ABC.

Thus ΠABF = ΠFBC = 1/2 ΠABC

Polygon

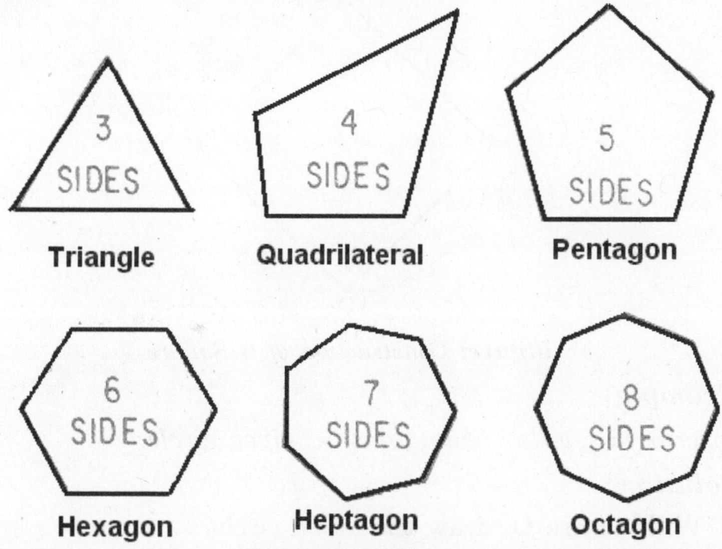

Figure: *Polygons*

A polygon is a plane figure bounded by three or more straight sides. A regular polygon has all its sides and angles equal. Specific polygons are named according to number of sides.

(a) A *triangle* has three sides.

(b) A *quadrilateral* has four sides.

(c) A *pentagon* has five sides.

(d) A *hexagon* has six sides.

(e) A *heptagon* has seven sides.

(f) An *octagon* has eight sides.

(g) A *nonagon* has nine sides.

(h) A *decagon* has ten sides.

(j) A *dodecagon* has twelve sides.

Regular Polygons Inscribed in Circles

Example

Inscribe a square in a given circle.

Solution

(a) With centre O, draw the given circle.

(b) Draw diametres AB and CD perpendicular to each other.

(c) Draw lines AC, CB, BD and DA, thus completing the square.

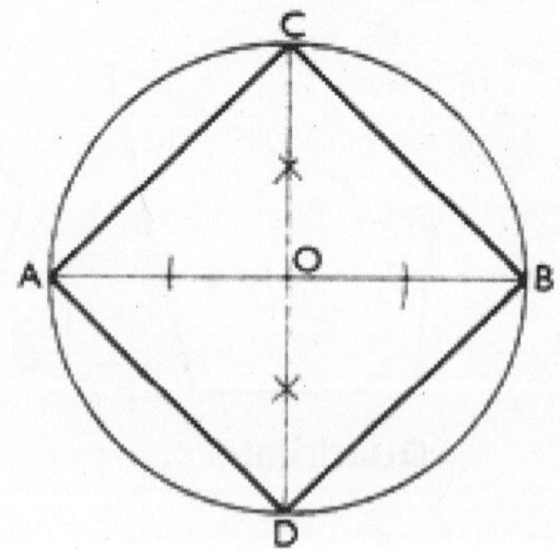

Figure: *Construction of a Square*

Example

Inscribe a regular pentagon in a given circle.

Solution

(a) With centre O, draw the given circle.

(b) Draw diametres AB and CD perpendicular to each other.

(c) Bisect AO in a point P. With centre P and radius PC, draw an arc cuttingOBin Q.

(d) With centre C and radius CQ, draw an arc cutting the circle in E and F.

(e) With centres E and F and the same radius, draw arcs cutting the circle in C and H respectively.

(f) Draw lines CE, EG, GH, HF and FC, thus completing the required pentagon

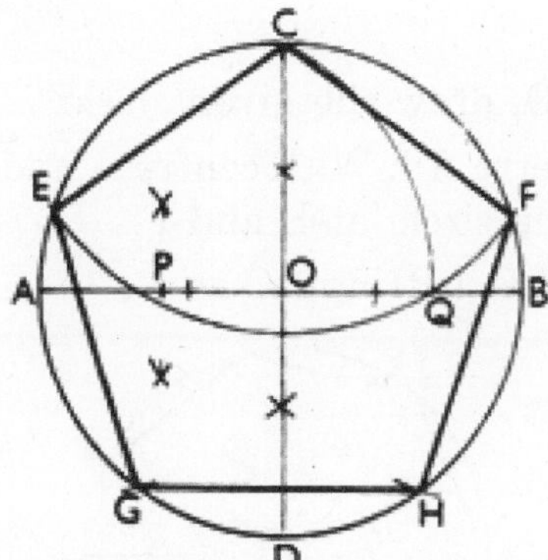

Figure: *Construction of a Regular Pentagon*

Example

Inscribe a regular hexagon in a given circle.

Solution

(a) With any point O as centre and radius equal to the given length, draw a circle.

(b) With the same radius and starting from any point on the circle, mark six equal divisions on the circle.

(c) Join the division points in proper sequence and complete the hexagon.

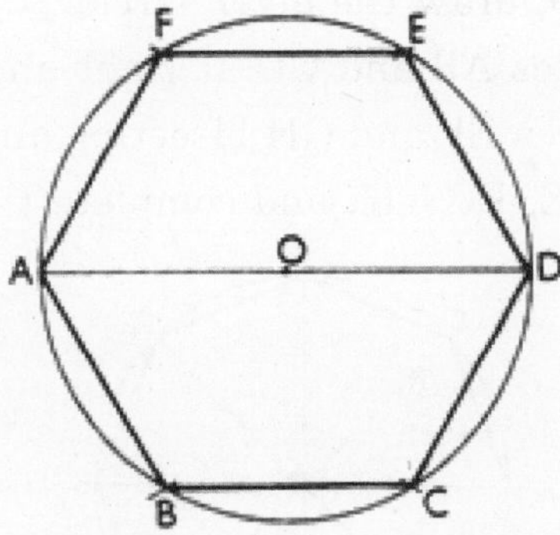

Figure: *Construction of a Regular Hexagon*

Note: When two sides of the hexagon are required to be horizontal the starting point for marking equal divisions should be on an end of the horizontal diametre.

Example

Inscribe a regular heptagon in a given circle.

Solution

(a) With centre O, draw the given circle.

(b) Draw a diametre AB. With centre A and radius AO, draw an arc cutting the circle at E and F.

(c) Draw a line EF, cutting AO in G.

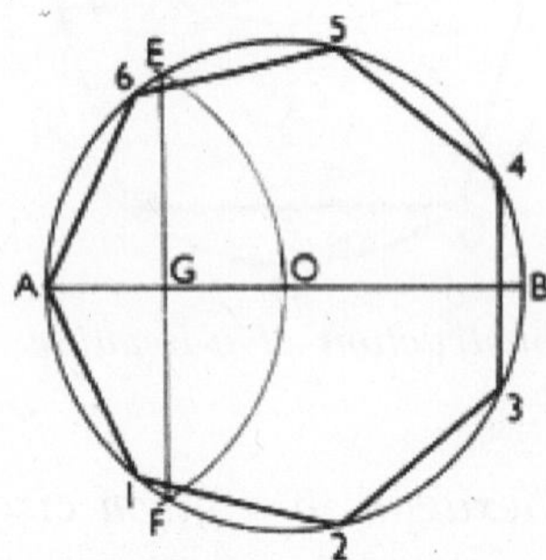

Figure: *Construction of a Regular Heptagon*

Then EG or FG is the length of the side of the heptagon. Therefore, from any point on the circle, say A, mark divisions equal to EG, around the circle. Join the division-points and obtain the heptagon.

Example

Inscribe a regular octagon in a given circle.

Solution

(a) With centre O, draw the given circle.

(b) Draw diametres AB and CD at right angles to each other.

(c) Draw diametres EF and GH bisecting angles AOC and COB.

(d) Draw lines AE, EC, etc. and complete the octagon.

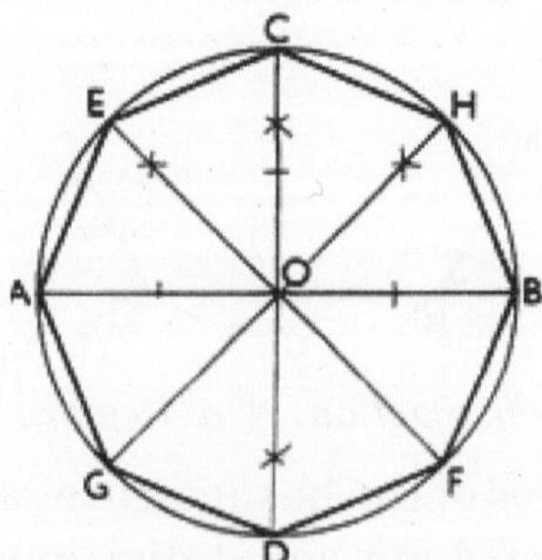

Figure: *Construction of a Regular Octagon*

Example

Construct a regular polygon, given the length of its side. Let the number of sides of the polygon be seven (i.e. heptagon).

Solution

General method for drawing any polygon:

(a) Draw a line AB equal to the given length.

(b) At B, draw a line BP perpendicular and equal to AB.

(c) Draw a line joining A with P.

(d) With centre B and radius AB, draw the quadrant AP.

(e) Draw the perpendicular bisector of AB to intersect the straight line AP at point 4 and the arc AP at 6.

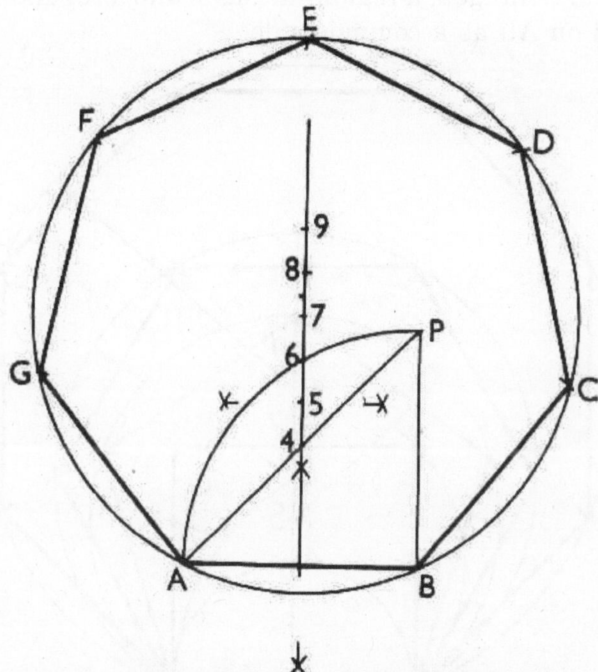

Figure: *Construction of a Regular Heptagon*

(f) Regular polygons of any number of sides can now be drawn as follows:

(i) A *square* of a side equal to AB can be inscribed in the circle drawn with centre 4 and radius A4.

(ii) A *regular hexagon* of a side equal to AB can be inscribed in the circle drawn with centre 6 and radius A6.

(iii) The mid-point 5 of the line 4-6 is the centre of the circle of the radius A5 in which a *regular pentagon* of a side equal to AB can be inscribed.

(iv) To locate centre 7 for the *regular heptagon* of side AB, mark a division 6-7 equal to the division 5-6.

(aa) With centre 7 and radius equal to A7, draw a circle.

(ab) Starting from B, cut it in seven equal divisions with radius equal to AB.

(ac) Draw lines BC, CD etc. and complete the heptagon.

Note: Regular polygons of any number of sides can be drawn by this method.

A regular pentagon, a regular hexagon and a regular octagon, all constructed on AB as a common side.

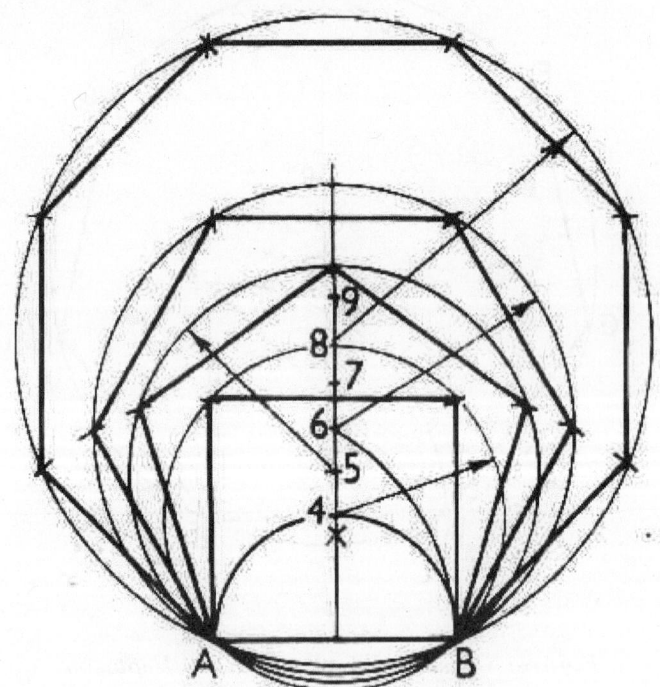

Figure: *Construction of Regular Polygons*

Bird's-eye View

A bird's-eye view is an elevated view of an object from above, with a perspective as though the observer were a bird, often used in the making of blueprints, floor plans and maps.

It can be an aerial photograph, but also a drawing. Before manned flight was common, the term "bird's eye" was used to distinguish views drawn from direct observation at high locations (for example a mountain or tower), from those constructed from an imagined (bird's) perspectives. Bird's eye views as a genre have existed since classical times. The last great flourishing of them was in the mid-to-late 19th century, when bird's eye view prints were popular in the United States and Europe.

The terms aerial view and aerial viewpoint are also sometimes used synonymous with bird's-eye view. The term *aerial view* can refer to any view from a great height, even at a wide angle, as for example when looking sideways from an airplane window or from a mountain top. Overhead view is fairly synonymous with *bird's-eye view* but tends to imply a less lofty vantage point than the latter term. For example, in computer and video games, an "overhead view" of a character or situation often places the vantage point only a few feet (a metre or two) above human height.

Recent technological and networking developments have made satellite images more accessible. Microsoft Bing Maps offers direct overhead satellite photos of the entire planet but also offers a feature named Bird's eye view in some locations.

The *Bird's Eye* photos are angled at 40 degrees rather than being straight down. Satellite imaging programmes and photos have been described as offering a viewer the opportunity to "fly over" and observe the world from this specific angle.

In filmmaking and video production, a bird's-eye shot refers to a shot looking directly down on the subject. The perspective is very foreshortened, making the subject appear short and squat. This shot can be used to give an overall establishing shot of a scene, or to emphasise the smallness or insignificance of the subjects. These shots are normally used for battle scenes or establishing where the character is. It is shot by lifting the camera up by hands or by hanging it off something strong enough to support it. When a scene needs a large area shot, it is a crane shot.

Centrolinead

The centrolinead was invented by Peter Nicholson, a British mathematician and architect, in 1814. It was used to construct 2-point perspective drawings where one or both vanishing points existed outside the drawing board. Draftsmen could use the instrument in pairs; one for each vanishing point on each side of the station point.

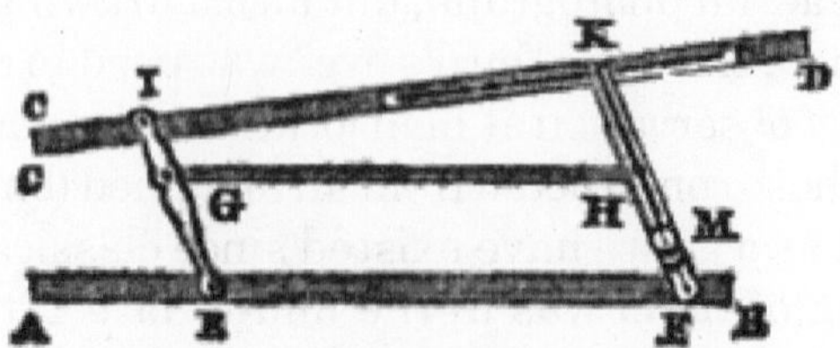

Figure: *Nicholson's centrolinead*

Centrolineads were produced in various sizes. Typically a brass fitting clamped the wooden arms together. Fittings were produced in both right and left-handed configuration, and certain adjustable designs could be used on either side.

Usage

Two short arms are set to form 90 degree angles against a third, longer drawing edge. Pins are placed near the edges of the drawing surface and serve as pivots for the arms. Pin placement is equidistant and symmetric across the horizontal line. A third centrolinead could be used to construct 3-point perspective.

Site Plan

A site plan is an architectural plan, landscape architecture document, and a detailed engineering drawing of proposed improvements to a given lot. A site plan usually shows a building footprint, travelways, parking, drainage facilities, sanitary sewer lines, water lines, trails, lighting, and landscaping and garden elements.

Such a plan of a site is a "graphic representation of the arrangement of buildings, parking, drives, landscaping and any other structure that is part of a development project".

A site plan is a "set of construction drawings that a builder or contractor uses to make improvements to a property. Counties can use the site plan to verify that development codes are being met and as a historical resource. Site plans are often prepared by a design consultant who must be either a licensed engineer, architect, landscape architect or land survey".

4

Compass and Straightedge

Construction

Compasses are usually made of metal or plastic, and consist of two parts connected by a hinge which can be adjusted to allow the changing of the radius of the circle drawn. Typically one part has a spike at its end, and the other part a pencil, or sometimes a pen.

Uses

Circles can be made by fastening one leg of the compasses into the paper with the spike, putting the pencil on the paper, and moving the pencil around while keeping the hinge on the same angle. The radius of the circle can be adjusted by changing the angle of the hinge.

Distances can be measured on a map using compasses with two spikes, also called a dividing compass. The hinge is set in such a way that the distance between the spikes on the map represents a certain distance in reality, and by measuring how many times the compasses fit between two points on the map the distance between those points can be calculated.

Oblique Pictorial

In an *oblique pictorial* drawing, the angles displayed among the axes, as well as the foreshortening factors (scale) are arbitrary. More precisely, any given set of three coplanar segments originating from the same point may be construed as forming some oblique perspective of three sides of a cube. This result is known as Pohlke's theorem, from the German mathematician Pohlke, who published it in the early 19th century.

The resulting distortions make the technique unsuitable for formal, working drawings. Nevertheless, the distortions are partially overcome by aligning one plane of the image parallel to the plane of projection. Doing so creates a true shape image of the chosen plane. This specific category of oblique projections, whereby lengths along the directions x and y are preserved, but lengths along direction z are drawn at angle using a reduction factor is very much in use for industrial drawings.

- *Cavalier projection* is the name of such a projection, where the length along the axis remains unscaled.
- *Cabinet projection*, popular in furniture illustrations, is an example of such a technique, wherein the receding axis is scaled to half-size (sometimes also two thirds the original).

Cavalier Projections

In cavalier projection (sometimes cavalier perspective or high view point) a point of the object is represented by three coordinates, x, y and z. On the drawing, it is represented by only two coordinates, x'' and y''. On the flat drawing, two axes, x and z on the figure, are perpendicular and the length on these axes are drawn with a 1:1 scale; it is thus similar to the dimetric projections, although it is not an orthographic projection, as the third axis, here y, is drawn in diagonal, making an arbitrary angle with the x'' axis, usually 30 or 45°. The length of the third axis is not scaled.

It is very easy to draw, especially with pen and paper. It is thus often used when a figure must be drawn by hand, e.g. on a black board (lesson, oral examination). The representation was initially used for military fortifications. In French, the « cavalier » (literally *rider, horseman*) is an artificial hill behind the walls that allows to see the enemy above the walls. The cavalier perspective was the way the things were seen from this high point. Some also explain the name by the fact that it was the way a rider could see a small object on the ground from his horseback.

Cabinet Projection

The term cabinet projection (sometimes cabinet perspective) stems from its use in illustrations by the furniture industry. Like cavalier perspective, one face of the projected object is parallel to the viewing plane, and the third axis is projected as going off in an angle (typical 30° or 45°). Unlike cavalier projection, where the third axis keeps its length, with cabinet projection the length of the receding lines is cut in half.

Mathematical Formula

As formula, if the plane facing the viewer is xy, and the receding axis is z, then a point P is projected like this:

$$P\begin{pmatrix} x \\ y \\ z \end{pmatrix} = \begin{pmatrix} x + 0.5 \cdot z \cdot \cos\alpha \\ y + 0.5 \cdot z \cdot \sin\alpha \\ 0 \end{pmatrix}$$

Where is the mentioned angle.

The transformation matrix is:

$$P = \begin{bmatrix} 1 & 0 & 0.5 \cdot \cos\alpha \\ 0 & 1 & 0.5 \cdot \sin\alpha \\ 0 & 0 & 0 \end{bmatrix}$$

Alternatively you could remove one third from the leading arm projected off the starting face, thus giving the same effect.

Perspective (Graphical)

Perspective in the graphic arts, such as drawing, is an approximate representation, on a flat surface (such as paper), of an image as it is seen by the eye. The two most characteristic features of perspective are that objects are drawn:

- Smaller as their distance from the observer increases
- Foreshortened: the size of an object's dimensions along the line of sight are relatively shorter than dimensions across the line of sight

Overview

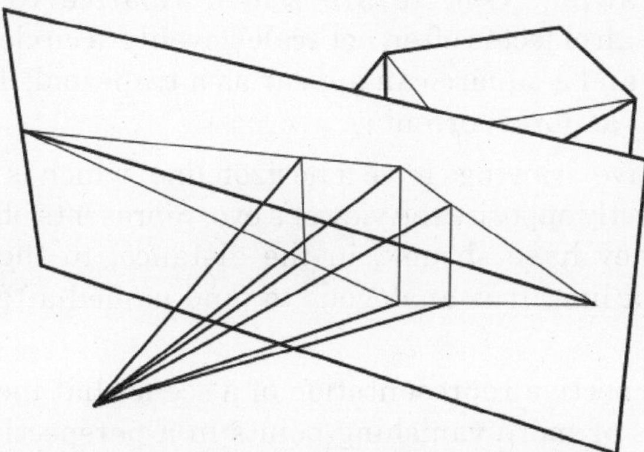

Figure: *A cube in two-point perspective.*

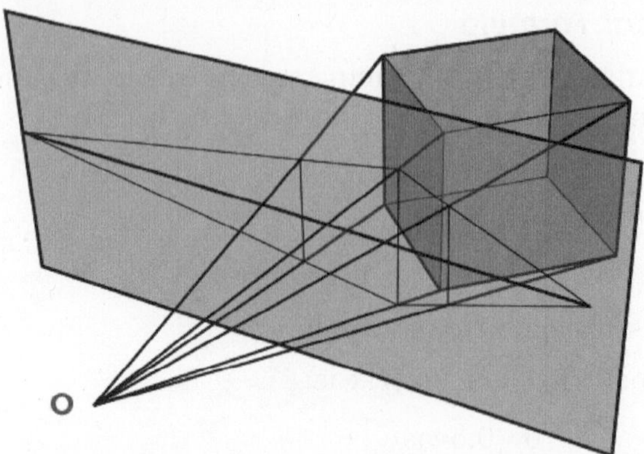

Figure: *Rays of light travel from the object, through the picture plane, and to the viewer's eye. This is the basis for graphical perspective.*

Linear perspective always works by representing the light that passes from a scene through an imaginary rectangle (the painting), to the viewer's eye. It is similar to a viewer looking through a window and painting what is seen directly onto the windowpane. If viewed from the same spot as the windowpane was painted, the painted image would be identical to what was seen through the unpainted window. Each painted object in the scene is a flat, scaled down version of the object on the other side of the window. Because each portion of the painted object lies on the straight line from the viewer's eye to the equivalent portion of the real object it represents, the viewer cannot perceive (sans depth perception) any difference between the painted scene on the windowpane and the view of the real scene. All perspective drawings assume the viewer is a certain distance away from the drawing. Objects are scaled relative to that viewer. Additionally, an object is often not scaled evenly: a circle often appears as an ellipse and a square can appear as a trapezoid. This distortion is referred to as foreshortening.

Perspective drawings have a horizon line, which is often implied. This line, directly opposite the viewer's eye, represents objects infinitely far away. They have shrunk, in the distance, to the infinitesimal thickness of a line. It is analogous to (and named after) the Earth's horizon.

Any perspective representation of a scene that includes parallel lines has one or more vanishing points in a perspective drawing. A one-point perspective drawing means that the drawing has a single vanishing point, usually (though not necessarily) directly opposite the viewer's eye and usually (though not necessarily) on the horizon line. All lines parallel with the viewer's line of sight recede to the horizon towards this vanishing point. This is the standard "receding railroad tracks" phenomenon. A two-point drawing would have lines parallel to two different angles. Any number of vanishing points are possible in a drawing, one for each set of parallel lines that are at an angle relative to the plane of the drawing.

Perspectives consisting of many parallel lines are observed most often when drawing architecture (architecture frequently uses lines parallel to the x, y, and z axes). Because it is rare to have a scene consisting solely of lines parallel to the three Cartesian axes (x, y, and z), it is rare to see perspectives in practice with only one, two, or three vanishing points; even a simple house frequently has a peaked roof which results in a minimum of six sets of parallel lines, in turn corresponding to up to six vanishing points.

In contrast, natural scenes often do not have any sets of parallel lines and thus no vanishing points.

Early History

The earliest art paintings and drawings typically sized objects and characters hierarchically according to their spiritual or thematic importance, not their distance from the viewer, and did not use foreshortening.

The most important figures are often shown as the highest in a composition, also from hieratic motives, leading to the "vertical perspective", common in the art of Ancient Egypt, where a group of "nearer" figures are shown below the larger figure or figures. The only method to indicate the relative position of elements in the composition was by overlapping, of which much use is made in works like the Parthenon Marbles.

Systematic attempts to evolve a system of perspective are usually considered to have begun around the fifth century BC. in the art of Ancient Greece, as part of a developing interest in illusionism allied to theatrical scenery and detailed within Aristotle's Poetics as 'skenographia':

Using flat panels on a stage to give the illusion of depth. The philosophers Anaxagoras and Democritus worked out geometric theories of perspective for use with skenographia. Alcibiades had paintings in his house designed based on skenographia, thus this art was not confined merely to the stage. Euclid's *Optics* introduced a mathematical theory of perspective; however, there is some debate over the extent to which Euclid's perspective coincides with a modern mathematical definition of perspective.

By the later periods of antiquity, artists, especially those in less popular traditions, were well aware that distant objects could be shown smaller than those close at hand for increased illusionism, but whether this convention was actually used in a work depended on many factors.

Some of the paintings found in the ruins of Pompeii show a remarkable realism and perspective for their time; it has been claimed that comprehensive systems of perspective were evolved in antiquity, but most scholars do not accept this. Hardly any of the many works where such a system would have been used have survived. A passage in Philostratus suggests that classical artists and theorists thought in terms of "circles" at equal distance from the viewer, like a classical semi-circular theatre seen from the stage.

The roof beams in rooms in the Vatican Virgil, from about 400 AD, are shown converging, more or less, on a common vanishing point, but this is not systematically related to the rest of the composition. In the Late Antique period use of perspective techniques declined.

The art of the new cultures of the Migration Period had no tradition of attempting compositions of large numbers of figures and Early Medieval art was slow and inconsistent in relearning the convention from classical models, though the process can be seen underway in Carolingian art.

European Medieval artists were aware of the general principle of varying the relative size of elements according to distance, but even more than classical art was perfectly ready to override it for other reasons. Buildings were often shown obliquely according to a particular convention. The use and sophistication of attempts to convey distance increased steadily during the period, but without a basis in a systematic theory. Byzantine art was also aware of these principles, but also had the reverse perspective convention for the setting of principal figures.

Giotto attempted drawings in perspective using an algebraic method to determine the placement of distant lines. The problem with using a linear ratio in this manner is that the apparent distance between a series of evenly spaced lines actually falls off with a sine dependence. To determine the ratio for each succeeding line, a recursive ratio must be used.

One of Giotto's first uses of his algebraic method of perspective was *Jesus Before Caiaphas.* Although the picture does not conform to the modern, geometrical method of perspective, it does give a considerable illusion of depth, and was a large step forward in Western art.

With the exception of dice, heraldry typically ignores perspective in the treatment of charges, though sometimes in later centuries charges are specified as *in perspective.*

Renaissance: Mathematical Basis

By the 14th century, Alhazen's *Book of Optics* was available in Italian translation, entitled *Deli Aspecti.* The Renaissance artist Lorenzo Ghiberti relied heavily upon this work, quoting it "verbatim and at length" while framing his account of art and its aesthetic imperatives in the "Commentario terzo." Alhazen's work was thus "central to the development of Ghiberti's thought about art and visual aesthetics" and "may well have been central to the development of artificial perspective in early Renaissance Italian painting."

In about 1413 a contemporary of Ghiberti, Filippo Brunelleschi, demonstrated the geometrical method of perspective, used today by artists, by painting the outlines of various Florentine buildings onto a mirror. When the building's outline was continued, he noticed that all of the lines converged on the horizon line. According to Vasari, he then set up a demonstration of his painting of the Baptistery in the incomplete doorway of the Duomo. He had the viewer look through a small hole on the back of the painting, facing the Baptistery. He would then set up a mirror, facing the viewer, which reflected his painting. To the viewer, the painting of the Baptistery and the Baptistery itself were nearly indistinguishable.

Soon after, nearly every artist in Florence and in Italy used geometrical perspective in their paintings, notably Masolino da Panicale and Donatello. Donatello started sculpting elaborate checkerboard floors into the simple manger portrayed in the birth of Christ. Although hardly historically accurate, these checkerboard floors obeyed the primary laws of geometrical perspective: the lines converged approximately to a vanishing point, and the rate at which the horizontal lines receded into the distance was graphically determined. This became an integral part of Quattrocento art. Melozzo da Forlì first used the technique of upward foreshortening (in Rome, Loreto, Forlì and others), and was celebrated for that. Not only was perspective a way of showing depth, it was also a new method of composing a painting. Paintings began to show a single, unified scene, rather than a combination of several.

As shown by the quick proliferation of accurate perspective paintings in Florence, Brunelleschi likely understood (with help from his friend the mathematician Toscanelli), but did not publish, the mathematics behind perspective. Decades later, his friend Leon Battista Alberti wrote *De pictura* (1435/1436), a treatise on proper methods of showing distance in painting. Alberti's primary breakthrough was not to show the mathematics in terms of conical projections, as it actually appears to the eye. Instead, he formulated the theory based on planar projections, or how the rays of light, passing from the viewer's eye to the landscape, would strike the picture plane (the painting). He was then able to calculate the apparent height of a distant object using two similar triangles.

The mathematics behind similar triangles is relatively simple, having been long ago formulated by Euclid. In viewing a wall, for instance, the first triangle has a vertex at the user's eye, and vertices at the top and bottom of the wall. The bottom of this triangle is the distance from the viewer to the wall.

The second, similar triangle, has a point at the viewer's eye, and has a length equal to the viewer's eye from the painting. The height of the second triangle can then be determined through a simple ratio, as proven by Euclid. Alberti was also trained in the science of optics through the school of Padua and under the influence of Biagio Pelacani da Parma who studied Alhazen's *Optics*.

Piero della Francesca elaborated on Della Pittura in his *De Prospectiva Pingendi* in the 1470s. Alberti had limited himself to figures on the ground plane and giving an overall basis for perspective. Della Francesca fleshed it out, explicitly covering solids in any area of the picture plane. Della Francesca also started the now common practice of using illustrated figures to explain the mathematical concepts, making his treatise easier to understand than Alberti's. Della Francesca was also the first to accurately draw the Platonic solids as they would appear in perspective.

Perspective remained, for a while, the domain of Florence. Jan van Eyck, among others, was unable to create a consistent structure for the converging lines in paintings, as in London's The Arnolfini Portrait, because he was unaware of the theoretical breakthrough just then occurring in Italy. However he achieved very subtle effects by manipulations of scale in his interiors. Gradually, and partly through the movement of academies of the arts, the Italian techniques became part of the training of artists across Europe, and later other parts of the world.

The culmination of these Renaissance traditions finds its ultimate synthesis in the research of the architect, geometre, and optician Girard Desargues on perspective, optics and projective geometry.

Present : Computer Graphics

3-D computer games and ray-tracers often use a modified version of perspective. Like the painter, the computer programme is generally not concerned with every ray of light that is in a scene. Instead, the programme simulates rays of light travelling backwards from the monitor (one for every pixel), and checks to see what it hits. In this way, the programme does not have to compute the trajectories of millions of rays of light that pass from a light source, hit an object, and miss the viewer.

CAD software, and some computer games (especially games using 3-D polygons) use linear algebra, and in particular matrix multiplication, to create a sense of perspective. The scene is a set of points, and these points are projected to a plane (computer screen) in front of the view point (the viewer's eye). The problem of perspective is simply finding the corresponding coordinates on the plane corresponding to the points in the scene. By the theories of linear algebra, a matrix multiplication directly computes the desired coordinates, thus bypassing any descriptive geometry theorems used in perspective drawing.

Types of Perspective

Of the many types of perspective drawings, the most common categorizations of artificial perspective are one-, two- and three-point. The names of these categories refer to the number of vanishing points in the perspective drawing.

One-point Perspective

A drawing has one-point perspective when it contains only one vanishing point on the horizon line. This type of perspective is typically used for images of roads, railway tracks, hallways, or buildings viewed so that the front is directly facing the viewer. Any objects that are made up of lines either directly parallel with the viewer's line of sight or directly perpendicular (the railroad slats) can be represented with one-point perspective. These parallel lines converge at the vanishing point.

One-point perspective exists when the painting plate (also known as the picture plane) is parallel to two axes of a rectilinear (or Cartesian) scene – a scene which is composed entirely of linear elements that intersect only at right angles. If one axis is parallel with the picture plane, then all elements are either parallel to the painting plate (either horizontally or vertically) or perpendicular to it. All elements that are parallel to the painting plate are drawn as parallel lines. All elements that are perpendicular to the painting plate converge at a single point (a vanishing point) on the horizon.

Two-point Perspective

A drawing has two-point perspective when it contains two vanishing points on the horizon line. In an illustration, these vanishing points can be placed arbitrarily along the horizon. Two-point perspective can be used to draw the same objects as one-point perspective, rotated: looking at the corner of a house, or looking at two forked roads shrink into the distance, for example. One point represents one set of parallel lines, the other point represents the other. Looking at a house from the corner, one wall would recede towards one vanishing point, the other wall would recede towards the opposite vanishing point.

Two-point perspective exists when the painting plate is parallel to a Cartesian scene in one axis (usually the z-axis) but not to the other two axes. If the scene being viewed consists solely of a cylinder sitting on a horizontal plane, no difference exists in the image of the cylinder between a one-point and two-point perspective.

Two-point perspective has one set of lines parallel to the picture plane and two sets oblique to it. Parallel lines oblique to the picture plane converge to a vanishing point,which means that this set-up will require two vanishing points.

Three-point Perspective

Three-point perspective is usually used for buildings seen from above (or below). In addition to the two vanishing points from before, one for each wall, there is now one for how those walls recede into the ground. This third vanishing point will be below the ground. Looking up at a tall building is another common example of the third vanishing point. This time the third vanishing point is high in space.

Three-point perspective exists when the perspective is a view of a Cartesian scene where the picture plane is not parallel to any of the scene's three axes. Each of the three vanishing points corresponds with one of the three axes of the scene. One-point, two-point, and three-point perspectives appear to embody different forms of calculated perspective. The methods required to generate these perspectives by hand are different. Mathematically, however, all three are identical: The difference is simply in the relative orientation of the rectilinear scene to the viewer.

Four-point Perspective

Four-point perspective, also called infinite-point perspective, is the curvilinear variant of two-point perspective. As the result when made into an infinite point version (i.e. when the amount of vanishing points exceeds the minimum amount required), a four point perspective image becomes a panorama that can go to a 360 degree view and beyond – when going beyond the 360 degree view the artist might depict an "impossible" room as the artist might depict something new when it's supposed to show part of what already exists within those 360 degrees.

This elongated frame can be used both horizontally and vertically and when used vertically can be described as an image that depicts both a worm's- and bird's-eye view of a scene at the same time.

Like all other foreshortened variants of perspective (respectively one- to six-point perspective), it starts off with a horizon line, followed by four equally spaced vanishing points to delineate four vertical lines.

The vanishing points made to create the curvilinear orthogonals are thus made ad hoc on the four vertical lines placed on the opposite side of the horizon line. The only dimension not foreshortened in this type of perspective is the rectilinear and parallel lines perpendicular to the horizon line – similar to the vertical lines used in two-point perspective.

Zero-point Perspective

Because vanishing points exist only when parallel lines are present in the scene, a perspective with no vanishing points ("zero-point" perspective) occurs if the viewer is observing a non-rectilinear scene. The most common example of a nonlinear scene is a natural scene (e.g., a mountain range) which frequently does not contain any parallel lines. A perspective without vanishing points can still create a sense of depth.

Other Varieties of Linear Perspective

One-point, two-point, and three-point perspective are dependent on the structure of the scene being viewed. These only exist for strict Cartesian (rectilinear) scenes.

By inserting into a Cartesian scene a set of parallel lines that are not parallel to any of the three axes of the scene, a new distinct vanishing point is created.

Therefore, it is possible to have an infinite-point perspective if the scene being viewed is not a Cartesian scene but instead consists of infinite pairs of parallel lines, where each pair is not parallel to any other pair. Zero point perspective view is also known as elevation of any object

Foreshortening

Foreshortening is the visual effect or optical illusion that causes an object or distance to appear shorter than it actually is because it is angled towards the viewer.

Although foreshortening is an important element in art where visual perspective is being depicted, foreshortening occurs in other types of two-dimensional representations of three-dimensional scenes. Some other types where foreshortening can occur include oblique parallel projection drawings. Foreshortening also occurs when imaging rugged terrain using a synthetic aperture radar system.

In painting, foreshortening in the depiction of the human figure was perfected in the Italian Renaissance, and the *The Lamentation over the Dead Christ* by Andrea Mantegna (1480s) is one of the most famous of a number of works that show off the new technique, which thereafter became a standard part of the training of artists.

Methods of Construction

Several methods of constructing perspectives exist, including:

- Freehand sketching (common in art)
- Graphically constructing (once common in architecture)
- Using a perspective grid
- Computing a perspective transform (common in 3D computer applications)
- Mimicry using tools such as a proportional divider (sometimes called a variscaler)

Example

One of the most common, and earliest, uses of geometrical perspective is a checkerboard floor. It is a simple but striking application of one-point perspective. Many of the properties of perspective drawing are used while drawing a checkerboard. The checkerboard floor is, essentially, just a combination of a series of squares. Once a single square is drawn, it can be widened or subdivided into a checkerboard. Where necessary, lines and points will be referred to by their colours in the diagram.

To draw a square in perspective, the artist starts by drawing a horizon line (*black*) and determining where the vanishing point (*green*) should be. The higher up the horizon line is, the lower the viewer will appear to be looking, and vice versa. The more off-centre the vanishing point, the more tilted the square will be. Because the square is made up of right angles, the vanishing point should be directly in the middle of the horizon line. A rotated square is drawn using two-point perspective, with each set of parallel lines leading to a different vanishing point.

The foremost edge of the (*orange*) square is drawn near the bottom of the painting. Because the viewer's picture plane is parallel to the bottom of the square, this line is horizontal. Lines connecting each side of the foremost edge to the vanishing point are drawn (in grey). These lines give the basic, one point "railroad tracks" perspective. The closer it is the horizon line, the farther away it is from the viewer, and the smaller it will appear. The farther away from the viewer it is, the closer it is to being perpendicular to the picture plane.

A new point (*the eye*) is now chosen, on the horizon line, either to the left or right of the vanishing point. The distance from this point to the vanishing point represents the distance of the viewer from the drawing. If this point is very far from the vanishing point, the square will appear squashed, and far away. If it is close, it will appear stretched out, as if it is very close to the viewer.

A line connecting this point to the opposite corner of the square is drawn. Where this (blue) line hits the side of the square, a horizontal line is drawn, representing the farthest edge of the square. The line just drawn represents the ray of light travelling from the farthest edge of the square to the viewer's eye. This step is key to understanding perspective drawing. The light that passes through the picture plane obviously can not be traced. Instead, lines that represent those rays of light are drawn on the picture plane. In the case of the square, the side of the square also represents the picture plane (at an angle), so there is a small shortcut: when the line hits the side of the square, it has also hit the appropriate spot in the picture plane. The (*blue*) line is drawn to the opposite edge of the foremost edge because of another shortcut: since all sides are the same length, the foremost edge can stand in for the side edge.

Original formulations used, instead of the side of the square, a vertical line to one side, representing the picture plane. Each line drawn through this plane was identical to the line of sight from the viewer's eye to the drawing, only rotated around the y-axis ninety degrees. It is, conceptually, an easier way of thinking of perspective. It can be easily shown that both methods are mathematically identical, and result in the same placement of the farthest side.

Limitations

Plato was one of the first to discuss the problems of perspective. "Thus (through perspective) every sort of confusion is revealed within us; and this is that weakness of the human mind on which the art of conjuring and of deceiving by light and shadow and other ingenious devices imposes, having an effect upon us like magic... And the arts of measuring and numbering and weighing come to the rescue of the human understanding – there is the beauty of them – and the apparent greater or less, or more or heavier, no longer have the mastery over us, but give way before calculation and measure and weight?"

Perspective images are calculated assuming a particular vanishing point. In order for the resulting image to appear identical to the original scene, a viewer of the perspective must view the image from the exact vantage point used in the calculations relative to the image. This cancels out what would appear to be distortions in the image when viewed from a different point.

These apparent distortions are more pronounced away from the centre of the image as the angle between a projected ray (from the scene to the eye) becomes more acute relative to the picture plane. In practice, unless the viewer chooses an extreme angle, like looking at it from the bottom corner of the window, the perspective normally looks more or less correct. This is referred to as "Zeeman's Paradox." It has been suggested that a drawing in perspective still seems to be in perspective at other spots because we still perceive it as a drawing, because it lacks depth of field cues.

For a typical perspective, however, the field of view is narrow enough (often only 60 degrees) that the distortions are similarly minimal enough that the image can be viewed from a point other than the actual calculated vantage point without appearing significantly distorted. When a larger angle of view is required, the standard method of projecting rays onto a flat picture plane becomes impractical. As a theoretical maximum, the field of view of a flat picture plane must be less than 180 degrees (as the field of view increases towards 180 degrees, the required breadth of the picture plane approaches infinity).

In order to create a projected ray image with a large field of view, one can project the image onto a curved surface. In order to have a large field of view horizontally in the image, a surface that is a vertical cylinder (i.e., the axis of the cylinder is parallel to the z-axis) will suffice (similarly, if the desired large field of view is only in the vertical direction of the image, a horizontal cylinder will suffice). A cylindrical picture surface will allow for a projected ray image up to a full 360 degrees in either the horizontal or vertical dimension of the perspective image (depending on the orientation of the cylinder). In the same way, by using a spherical picture surface, the field of view can be a full 360 degrees in any direction (note that for a spherical surface, all projected rays from the scene to the eye intersect the surface at a right angle).

Just as a standard perspective image must be viewed from the calculated vantage point for the image to appear identical to the true scene, a projected image onto a cylinder or sphere must likewise be viewed from the calculated vantage point for it to be precisely identical to the original scene. If an image projected onto a cylindrical surface is "unrolled" into a flat image, different types of distortions occur. For example, many of the scene's straight lines will be drawn as curves. An image projected onto a spherical surface can be flattened in various ways:

- An image equivalent to an unrolled cylinder
- A portion of the sphere can be flattened into an image equivalent to a standard perspective
- An image similar to a fisheye photograph

Curvilinear Perspective

Curvilinear perspective is a graphical projection used to draw 3D objects on 2D surfaces. It was formally codified in 1968 by the artists and art historians André Barre and Albert Flocon in the book *La*

Perspective curviligne, which was translated into English in 1987 as *Curvilinear Perspective: From Visual Space to the Constructed Image* and published by the University of California Press.

Background

In 1959, Flocon had acquired a copy of *Grafiek en tekeningen* by M. C. Escher who strongly impressed him with his use of bent and curved perspective, which influenced the theory Flocon and Barre were developing. They started a long correspondence, in which Escher called Flocon a "kindred spirit".

Horizon and Vanishing Points

The system uses curving perspective lines instead of straight converging ones to approximate the image on the retina of the eye, which is itself spherical, more accurately than the traditional linear perspective, which uses straight lines and gets very strangely distorted at the edges.

It uses either four or five vanishing points:

- In five-point (fisheye) perspective: Four vanishing points are placed around in a circle, they are named N, W, S, E, and one vanishing point in the centre of the circle.
- Four, or infinite-point perspective is the one that (arguably) most approximates the perspective of the human eye, while at the same time being effective for making impossible spaces, while five point is the curvilinear equivalent of one point perspective, so is four point the equivalent of two point perspective.

This technique can, like two-point perspective, use a vertical line as a horizon line, creating both a worms and birds eye view at the same time. It uses four or more points equally spaced along a horizon line, all vertical lines are made perpendicular to the horizon line, while orthogonals are created using a compass set on a line made at a 90-degree angle through each of the four vanishing points.

History

Earlier, less mathematically precise versions can be seen in the work of the miniaturist Jean Fouquet. Leonardo da Vinci in a lost notebook spoke of curved perspective lines.

Examples of approximated (not necessarily systematically constructed, but emulated through an empirical method) five-point perspective can also be found in several mannerist paintings such as the curved mirror in Jan van Eyck's *Arnolfini's Wedding* as well as in the famous self-portrait of Parmigianino seen through a shaving mirror.

The book *Vanishing Point: Perspective for Comics from the Ground Up* by Jason Cheeseman-Meyer teaches five and four (infinite) point perspective.

Geometric Relationship

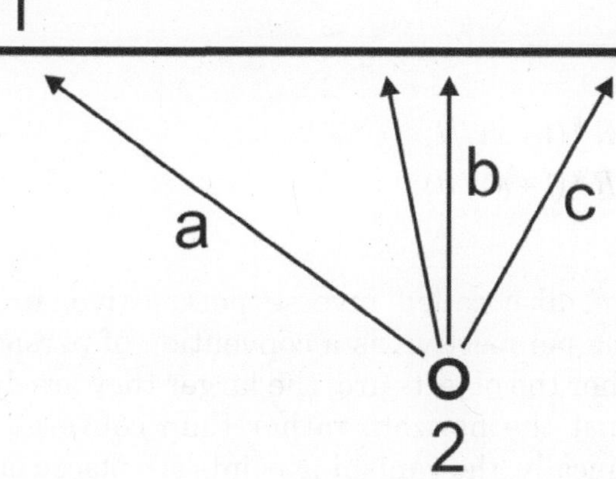

Figure: *shows the wall* 1 *and the observer* 2 *from the upper projection*

Distances *a* and *c* between the viewer and the wall are greater than the *b* distance , so adopting the principle that when an object is a greater distance from the observer, it becomes smaller, the wall should be reduced and thus becomes distorted at the edges.

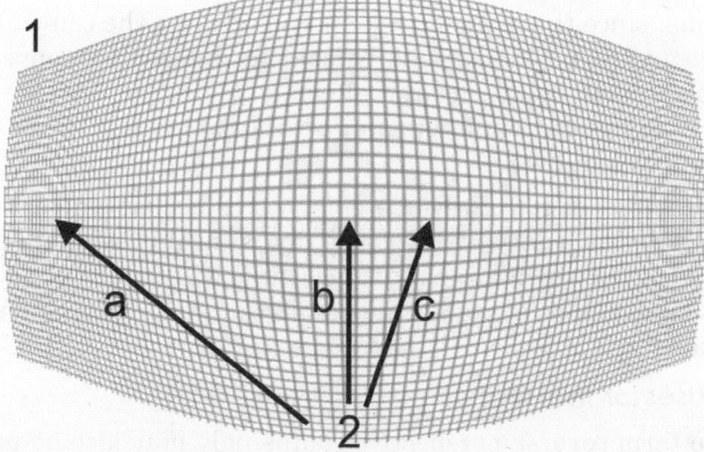

Figure: *shows the same situation from the observer's view point.*

Mathematics

If a point has the 3D Cartesian coordinates (x_1, y_1, z_1):

$$Pt3D = \begin{cases} x1 \\ y1 \\ z1 \end{cases}$$

the transformation of this point to a curvilinear reference system of radius R is:

$$d = \sqrt{x1^2 + y1^2 + z1^2}$$

$$Pt2D = \begin{cases} x = R * (1 + x1 / d) \\ y = R * (1 + y1 / d) \end{cases}.$$

Reverse Perspective

Reverse perspective, also called inverse perspective, inverted perspective or Byzantine perspective, is a convention of perspective drawing where the further the objects are, the larger they are drawn. The lines diverge against the horizon, rather than converge as in linear perspective. Technically, the vanishing points are placed outside the painting with the illusion that they are "in front of" the painting.

The name Byzantine perspective comes from the use of this perspective in Byzantine and Russian Orthodox icons; it is also found in East Asian art, and was sometimes used in Cubism and other movements of modern art.

The reasons for the convention are still debated among art historians; since the artists concerned in forming the convention did not have access to the more realistic linear perspective convention it is not clear how deliberate the effects achieved were. The scheme shows the image content as opening up and expanding, increasing the viewer's sense of awe.

A practical effect of reverse perspective is that since our vision has its greatest acuity at the focus, a visual representation which enlarges things which are not at the centre will tend to even out the lack of discernment of detail, thus aiding in the envisionment of the image as a gestalt.

Other Uses of the Term

The term reverse perspective confusingly may also be used of a rather different convention, found in the Art of ancient Egypt and other cultures, and also known as vertical perspective, where minor scenes at the bottom of a painted image or carved relief are shown at a far smaller scale than the main figures higher up. This system should not be understood as an attempt to convey any visual reality in the connection between the sections at different scales, whose size is dictated by their relative importance.

The term reverse perspective can also refer to the work of Patrick Hughes. Reverse perspective in this context refers to the illusion that is created when the two dimensional perspective of a painting is reversed by the three dimensional relief on which it is painted. The effect is to confuse the brain such that a completely false image is created that appears "to move" with the viewer.

3D Projection

3D projection is any method of mapping three-dimensional points to a two-dimensional plane. As most current methods for displaying graphical data are based on planar two-dimensional media, the use of this type of projection is widespread, especially in computer graphics, engineering and drafting.

Orthographic Projection

When the human eye looks at a scene, objects in the distance appear smaller than objects close by. Orthographic projection ignores this effect to allow the creation of to-scale drawings for construction and engineering.

Orthographic projections are a small set of transforms often used to show profile, detail or precise measurements of a three dimensional object. Common names for orthographic projections include plane, cross-section, bird's-eye, and elevation.

If the normal of the viewing plane (the camera direction) is parallel to one of the primary axes (which is the *x*, *y*, or *z* axis), the mathematical transformation is as follows; To project the 3D point a_x, a_y, a_z onto the 2D point b_x, b_y using an orthographic projection parallel to the y axis (profile view), the following equations can be used:

$$b_x = s_x a_x + c_x$$

$$b_y = s_z a_z + c_z$$

where the vector s is an arbitrary scale factor, and c is an arbitrary offset. These constants are optional, and can be used to properly align the viewport. Using matrix multiplication, the equations become:

$$\begin{bmatrix} b_x \\ b_y \end{bmatrix} = \begin{bmatrix} s_x & 0 & 0 \\ 0 & 0 & s_z \end{bmatrix} \begin{bmatrix} a_x \\ a_y \\ a_z \end{bmatrix} + \begin{bmatrix} c_x \\ c_z \end{bmatrix}.$$

While orthographically projected images represent the three dimensional nature of the object projected, they do not represent the object as it would be recorded photographically or perceived by a viewer observing it directly. In particular, parallel lengths at all points in an orthographically projected image are of the same scale regardless of whether they are far away or near to the virtual viewer. As a result, lengths near to the viewer are not foreshortened as they would be in a perspective projection.

Weak Perspective Projection

A "weak" perspective projection uses the same principles of an orthographic projection, but requires the scaling factor to be specified, thus ensuring that closer objects appear bigger in the projection, and vice-versa. It can be seen as a hybrid between an orthographic and a perspective projection, and described either as a perspective projection with individual point depths Z_i replaced by an average constant depth Z_{ave}, or simply as an orthographic projection plus a scaling.

The weak-perspective model thus approximates perspective projection while using a simpler model, similar to the pure (unscaled) orthographic perspective.

It is a reasonable approximation when the depth of the object along the line of sight is small compared to the distance Z_{ave} from the camera, and the field of view is small. With these conditions, it can be assumed that all points on a 3D object are at the same distance from the camera without significant errors in the projection (compared to the full perspective model).

Perspective Projection

When the human eye views a scene, objects in the distance appear smaller than objects close by - this is known as perspective. While orthographic projection ignores this effect to allow accurate measurements, perspective definition shows distant objects as smaller to provide additional realism.

The perspective projection requires a more involved definition as compared to orthographic projections. A conceptual aid to understanding the mechanics of this projection is to imagine the 2D projection as though the object(s) are being viewed through a camera viewfinder. The camera's position, orientation, and field of view control the behaviour of the projection transformation. The following variables are defined to describe this transformation:

- $a_{x,y,z}$ - the 3D position of a point *A* that is to be projected.
- $c_{x,y,z}$ - the 3D position of a point *C* representing the camera.
- $\theta_{x,y,z}$ - The orientation of the camera (represented, for instance, by Tait–Bryan angles).
- $e_{x,y,z}$ - the viewer's position relative to the display surface.

Which results in:

- $b_{x,y}$ - the 2D projection of *a*.

When $c_{x,y,z} = \langle 0,0,0 \rangle$, and $\theta_{x,y,z} = \langle 0,0,0 \rangle$, the 3D vector $\langle 1,2,0 \rangle$ is projected to the 2D vector $\langle 1,2 \rangle$.

Otherwise, to compute $b_{x,y}$ we first define a vector $d_{x,y,z}$ as the position of point *A* with respect to a coordinate system defined by the camera, with origin in *C* and rotated by θ with respect to the initial coordinate system.

This is achieved by subtracting c from a and then applying a rotation by $-\theta$ to the result. This transformation is often called a camera transform, and can be expressed as follows, expressing the rotation in terms of rotations about the *x*, *y*, and *z* axes (these calculations assume that the axes are ordered as a left-handed system of axes):

$$\begin{bmatrix} d_x \\ d_y \\ d_z \end{bmatrix} = \begin{bmatrix} 1 & 0 & 0 \\ 0 & \cos\theta_x & -\sin\theta_x \\ 0 & \sin\theta_x & \cos\theta_x \end{bmatrix} \begin{bmatrix} \cos\theta_y & 0 & \sin\theta_y \\ 0 & 1 & 0 \\ -\sin\theta_y & 0 & \cos\theta_y \end{bmatrix} \begin{bmatrix} \cos\theta_z & -\sin\theta_z & 0 \\ \sin\theta_z & \cos\theta_z & 0 \\ 0 & 0 & 1 \end{bmatrix} \left(\begin{bmatrix} a_x \\ a_y \\ a_z \end{bmatrix} - \begin{bmatrix} c_x \\ c_y \\ c_z \end{bmatrix} \right)$$

This representation corresponds to rotating by three Euler angles (more properly, Tait–Bryan angles), using the *xyz* convention, which can be interpreted either as "rotate about the *extrinsic* axes (axes of the *scene*) in the order *z, y, x* (reading right-to-left)" or "rotate about the *intrinsic* axes (axes of the *camera*) in the order *x, y, z* (reading left-to-right)". Note that if the camera is not rotated ($\theta_{x,y,z} = \langle 0,0,0 \rangle$), then the matrices drop out (as identities), and this reduces to simply a shift: $d = a - c$.

Alternatively, without using matrices, (note that the signs of angles are inconsistent with matrix form):

$$\begin{aligned} d_x &= \cos\theta_y \cdot (\sin\theta_z \cdot (a_y - c_y) + \cos\theta_z \cdot (a_x - c_x)) - \sin\theta_y \cdot (a_z - c_z) \\ d_y &= \sin\theta_x \cdot (\cos\theta_y \cdot (a_z - c_z) + \sin\theta_y \cdot (\sin\theta_z \cdot (a_y - c_y) + \cos\theta_z \cdot (a_x - c_x))) + \cos\theta_x \cdot (\cos\theta_z \cdot (a_y - c_y) - \sin\theta_z \cdot (a_x - c_x)) \\ d_z &= \cos\theta_x \cdot (\cos\theta_y \cdot (a_z - c_z) + \sin\theta_y \cdot (\sin\theta_z \cdot (a_y - c_y) + \cos\theta_z \cdot (a_x - c_x))) - \sin\theta_x \cdot (\cos\theta_z \cdot (a_y - c_y) - \sin\theta_z \cdot (a_x - c_x)) \end{aligned}$$

This transformed point can then be projected onto the 2D plane using the formula (here, x/y is used as the projection plane; literature also may use x/z):

$$\begin{aligned} b_x &= \frac{e_z}{d_z} d_x - e_x \\ b_y &= \frac{e_z}{d_z} d_y - e_y \end{aligned}$$

Or, in matrix form using homogeneous coordinates, the system

$$\begin{bmatrix} f_x \\ f_y \\ f_z \\ f_w \end{bmatrix} = \begin{bmatrix} 1 & 0 & -\frac{e_x}{e_z} & 0 \\ 0 & 1 & -\frac{e_y}{e_z} & 0 \\ 0 & 0 & 1 & 0 \\ 0 & 0 & 1/e_z & 0 \end{bmatrix} \begin{bmatrix} d_x \\ d_y \\ d_z \\ 1 \end{bmatrix}$$

in conjunction with an argument using similar triangles, leads to division by the homogeneous coordinate, giving

$$\begin{aligned} b_x &= f_x / f_w \\ b_y &= f_y / f_w \end{aligned}$$

The distance of the viewer from the display surface, e_z, directly relates to the field of view, where $\alpha = 2 \tan^{-1}(1/e_z)$ is the viewed angle. (Note: This assumes that you map the points (-1,-1) and (1,1) to the corners of your viewing surface)

The above equations can also be rewritten as:

$$\begin{aligned} b_x &= (d_x s_x)/(d_z r_x) r_z \\ b_y &= (d_y s_y)/(d_z r_y) r_z \end{aligned}$$

In which $s_{x,y}$ is the display size, $r_{x,y}$ is the recording surface size (CCD or film), r_z is the distance from the recording surface to the entrance pupil (camera centre), and d_z is the distance, from the 3D point being projected, to the entrance pupil.

Subsequent clipping and scaling operations may be necessary to map the 2D plane onto any particular display media.

Diagram

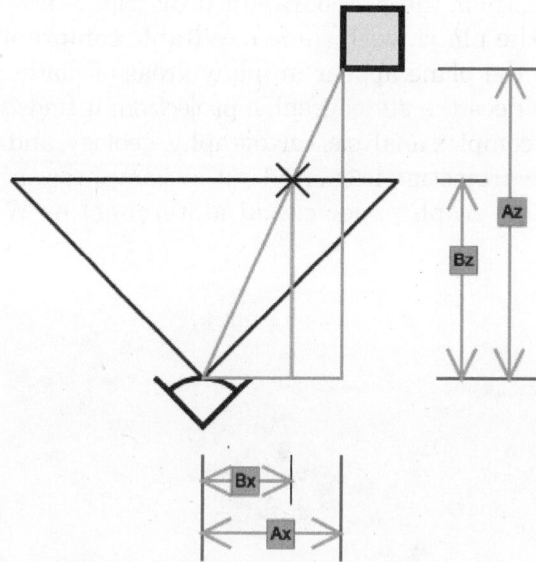

To determine which screen *x*-coordinate corresponds to a point at A_x, A_z multiply the point coordinates by:

$$B_x = A_x \frac{B_z}{A_z}$$

where

B_x is the screen *x* coordinate

A_x is the model *x* coordinate

B_z is the focal length—the axial distance from the camera centre to the image plane

A_z is the subject distance.

Because the camera is in 3D, the same works for the screen *y*-coordinate, substituting *y* for *x* in the above diagram and equation.

Stereographic Projection

In geometry, the stereographic projection is a particular mapping (function) that projects a sphere onto a plane. The projection is defined on the entire sphere, except at one point: the projection point. Where it is defined, the mapping is smooth and bijective. It is conformal, meaning that it preserves angles. It is neither isometric nor area-preserving: that is, it preserves neither distances nor the areas of figures.

Intuitively, then, the stereographic projection is a way of picturing the sphere as the plane, with some inevitable compromises. Because the sphere and the plane appear in many areas of mathematics and its applications, so does the stereographic projection; it finds use in diverse fields including complex analysis, cartography, geology, and photography. In practice, the projection is carried out by computer or by hand using a special kind of graph paper called a stereonet or Wulff net.

History

Figure: *Illustration by Rubens for "Opticorum libri sex philosophis juxta ac mathematicis utiles", by François d'Aiguillon. It demonstrates how the projection is computed.*

The stereographic projection was known to Hipparchus, Ptolemy and probably earlier to the Egyptians. It was originally known as the planisphere projection. *Planisphaerium* by Ptolemy is the oldest surviving document that describes it. One of its most important uses was the representation of celestial charts. The term *planisphere* is still used to refer to such charts.

It is believed that the earliest existing world map created by Gualterious Lud of St Dié, Lorraine, in 1507 is based upon the stereographic projection, mapping each hemisphere as a circular disk. The equatorial aspect of the stereographic projection, commonly used for maps of the Eastern and Western Hemispheres in the 17th and 18th centuries (and 16th century - Jean Roze 1542; Rumold Mercator 1595), was utilised by the ancient astronomers like Ptolemy

François d'Aiguillon gave the stereographic projection its current name in his 1613 work *Opticorum libri sex philosophis juxta ac mathematicis utiles* (Six Books of Optics, useful for philosophers and mathematicians alike). In 1695, Edmond Halley, motivated by his interest in star charts, published the first mathematical proof that this map is conformal. He used the recently established tools of calculus, invented by his friend Isaac Newton.

Definition

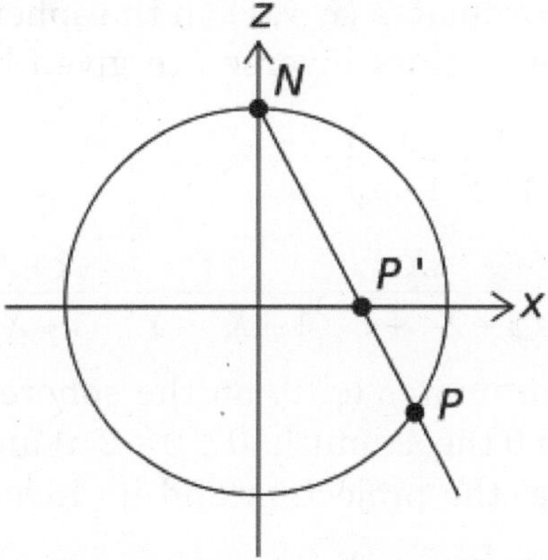

Figure: *Stereographic projection of the unit sphere from the north pole onto the plane z = 0, shown here in cross section*

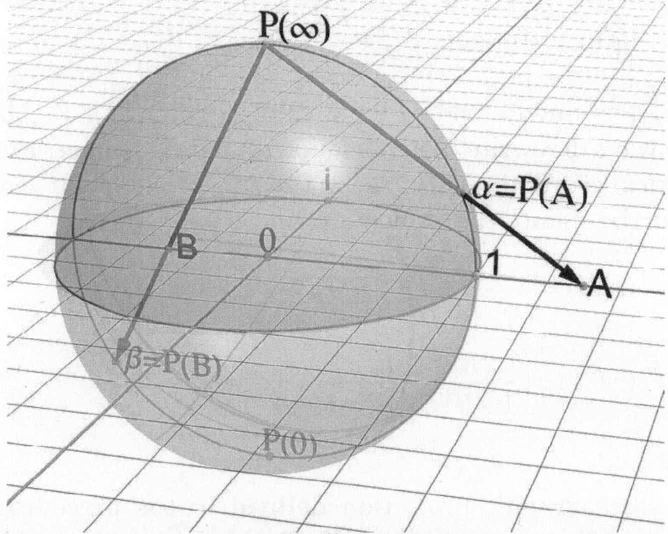

Figure: *This section focuses on the projection of the unit sphere from the north pole onto the plane through the equator. Other formulations are treated in later sections.*

The unit sphere in three-dimensional space R^3 is the set of points (x, y, z) such that $x^2 + y^2 + z^2 = 1$. Let $N = (0, 0, 1)$ be the "north pole", and let M be the rest of the sphere. The plane $z = 0$ runs through the centre of the sphere; the "equator" is the intersection of the sphere with this plane.

For any point P on M, there is a unique line through N and P, and this line intersects the plane $z = 0$ in exactly one point P'. Define the stereographic projection of P to be this point P' in the plane.

In Cartesian coordinates (x, y, z) on the sphere and (X, Y) on the plane, the projection and its inverse are given by the formulas

$$(X,Y) = \left(\frac{x}{1-z}, \frac{y}{1-z}\right),$$

$$(x,y,z) = \left(\frac{2X}{1+X^2+Y^2}, \frac{2Y}{1+X^2+Y^2}, \frac{-1+X^2+Y^2}{1+X^2+Y^2}\right).$$

In spherical coordinates (φ, θ) on the sphere (with φ the zenith angle, $0 \le \varphi \le \pi$, and θ the azimuth, $0 \le \theta \le 2\pi$) and polar coordinates (R, Θ) on the plane, the projection and its inverse are

$$(R,\Theta) = \left(\frac{\sin\varphi}{1-\cos\varphi}, \theta\right),$$

$$(\varphi,\theta) = \left(2\arctan\left(\frac{1}{R}\right), \Theta\right).$$

Here, φ is understood to have value π when $R = 0$. Also, there are many ways to rewrite these formulas using trigonometric identities. In cylindrical coordinates (r, θ, z) on the sphere and polar coordinates (R, Θ) on the plane, the projection and its inverse are

$$(R,\Theta) = \left(\frac{r}{1-z}, \theta\right),$$

$$(r,\theta,z) = \left(\frac{2R}{1+R^2}, \Theta, \frac{R^2-1}{R^2+1}\right).$$

Properties

The stereographic projection defined in the preceding section sends the "south pole" $(0, 0, -1)$ to $(0, 0)$, the equator to the unit circle, the southern hemisphere to the region inside the circle, and the northern hemisphere to the region outside the circle.

The projection is not defined at the projection point $N = (0, 0, 1)$. Small neighbourhoods of this point are sent to subsets of the plane far away from (0, 0). The closer P is to (0, 0, 1), the more distant its image is from (0, 0) in the plane. For this reason it is common to speak of (0, 0, 1) as mapping to "infinity" in the plane, and of the sphere as completing the plane by adding a "point at infinity". This notion finds utility in projective geometry and complex analysis. On a merely topological level, it illustrates how the sphere is homeomorphic to the one point compactification of the plane. In Cartesian coordinates a point $P(x, y, z)$ on the sphere and its image $P2$ (X, Y) on the plane either both are rational points or none of them:

$$P \in \mathbb{Q}^3 \Leftrightarrow P' \in \mathbb{Q}^2$$

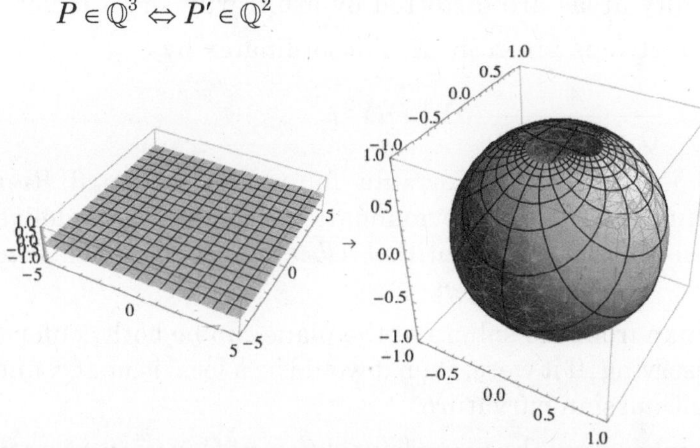

Figure: *A Cartesian grid on the plane appears distorted on the sphere. The grid lines are still perpendicular, but the areas of the grid squares shrink as they approach the north pole.*

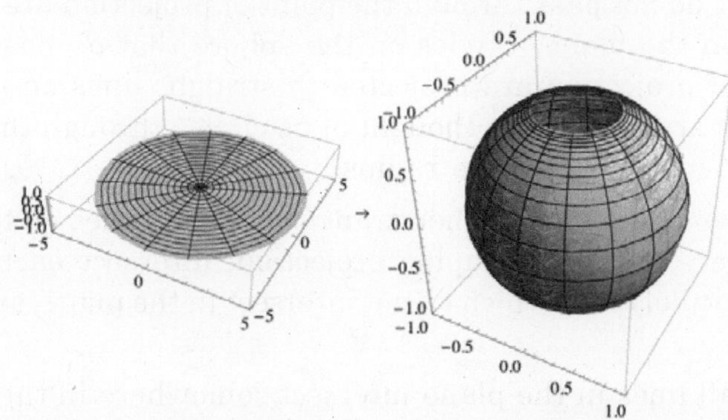

Figure: *A polar grid on the plane appears distorted on the sphere. The grid curves are still perpendicular, but the areas of the grid sectors shrink as they approach the north pole.*

Stereographic projection is conformal, meaning that it preserves the angles at which curves cross each other.

On the other hand, stereographic projection does not preserve area; in general, the area of a region of the sphere does not equal the area of its projection onto the plane. The area element is given in (X, Y) coordinates by

$$dA = \frac{4}{(1 + X^2 + Y^2)^2} \, dX \, dY.$$

Along the unit circle, where $X^2 + Y^2 = 1$, there is no infinitesimal distortion of area. Near (0, 0) areas are distorted by a factor of 4, and near infinity areas are distorted by arbitrarily small factors.

The metric is given in (X, Y) coordinates by

$$\frac{4}{(1 + X^2 + Y^2)^2} (dX^2 + dY^2),$$

and is the unique formula found in Bernhard Riemann's *Habilitationsschrift* on the foundations of geometry, delivered at Göttingen in 1854, and entitled *Über die Hypothesen welche der Geometrie zu Grunde liegen*.

No map from the sphere to the plane can be both conformal and area-preserving. If it were, then it would be a local isometry and would preserve Gaussian curvature.

The sphere and the plane have different Gaussian curvatures, so this is impossible. The conformality of the stereographic projection implies a number of convenient geometric properties. Circles on the sphere that do *not* pass through the point of projection are projected to circles on the plane. Circles on the sphere that *do* pass through the point of projection are projected to straight lines on the plane. These lines are sometimes thought of as circles through the point at infinity, or circles of infinite radius.

All lines in the plane, when transformed to circles on the sphere by the inverse of stereographic projection, intersect each other at infinity. Parallel lines, which do not intersect in the plane, are tangent at infinity.

Thus all lines in the plane intersect somewhere in the sphere—either transversally at two points, or tangently at infinity. (Similar remarks hold about the real projective plane, but the intersection relationships are different there.)

Figure: *The sphere, with various loxodromes shown in distinct colours*

The loxodromes of the sphere map to curves on the plane of the form

$$R = e^{\Theta/a},$$

where the parametre a measures the "tightness" of the loxodrome. Thus loxodromes correspond to logarithmic spirals. These spirals intersect radial lines in the plane at equal angles, just as the loxodromes intersect meridians on the sphere at equal angles.

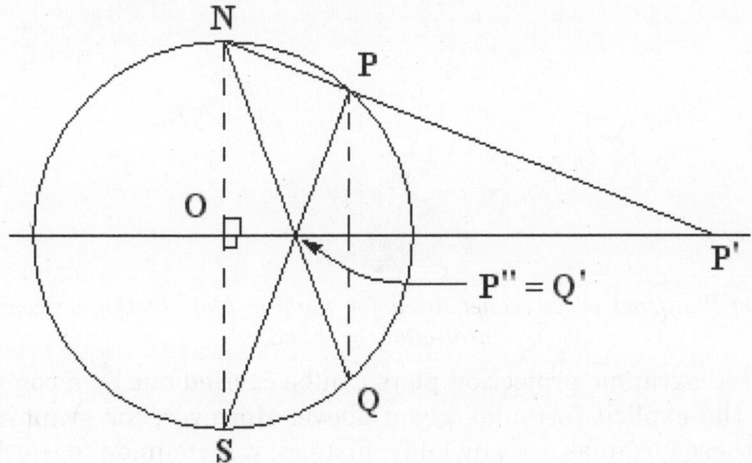

The stereographic projection relates to the plane inversion in a simple way. Let P and Q be two points on the sphere with projections P' and Q' on the plane. Then P' and Q' are inversive images of each other in the image of the equatorial circle if and only if P and Q are reflections of each other in the equatorial plane.

In other words, if:

- P is a point on the sphere, but not a 'north pole' N and not its antipode, the 'south pole' S,
- P' is the image of P in a stereographic projection with the projection point N and
- P" is the image of P in a stereographic projection with the projection point S,

then P' and P" are inversive images of each other in the unit circle.

$$\triangle NOP' \sim \triangle P''OS \Rightarrow OP' : ON = OS : OP'' \Rightarrow OP' \cdot OP'' = r^2$$

Wulff Net

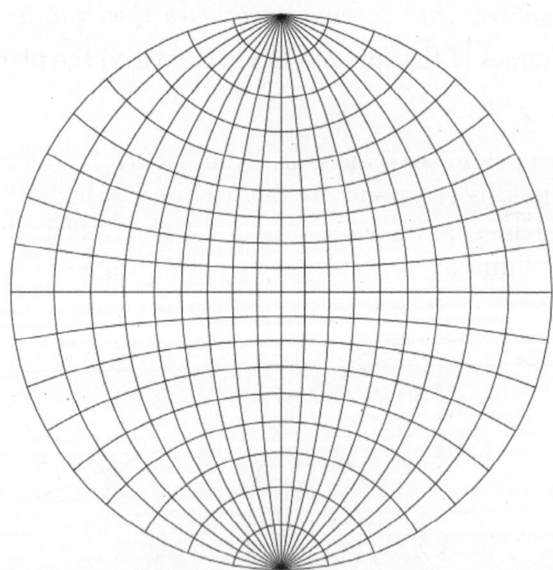

Figure: *Wulff net or stereonet, used for making plots of the stereographic projection by hand*

Stereographic projection plots can be carried out by a computer using the explicit formulas given above. However, for graphing by hand these formulas are unwieldy; instead, it is common to use graph paper designed specifically for the task. To make this graph paper, one places a grid of parallels and meridians on the hemisphere, and then stereographically projects these curves to the disk. The result is called a stereonet or Wulff net (named for the Russian mineralogist George (Yuri Viktorovich) Wulff).

In the figure, the area-distorting property of the stereographic projection can be seen by comparing a grid sector near the centre of the net with one at the far right of the net. The two sectors have equal areas on the sphere. On the disk, the latter has nearly four times the area as the former; if one uses finer and finer grids on the sphere, then the ratio of the areas approaches exactly 4.

On the Wulff net, the images of the parallels and meridians intersect at right angles. This orthogonality property is a consequence of the angle-preserving property of the stereoscopic projection. (However, the angle-preserving property is stronger than this property; not all projections that preserve the orthogonality of parallels and meridians are angle-preserving.)

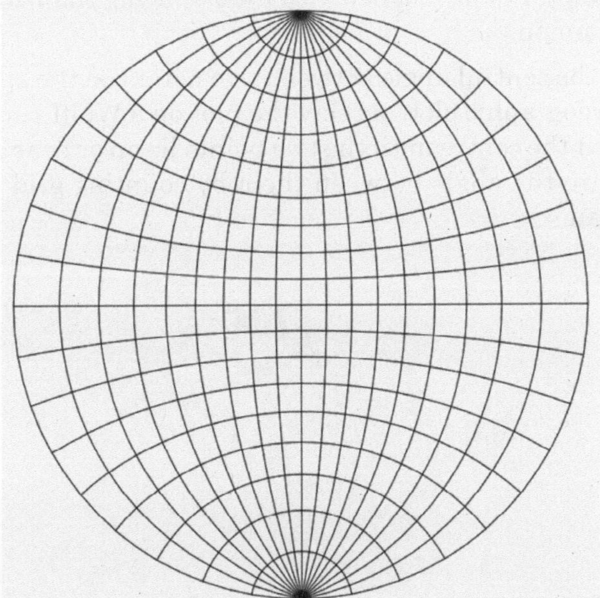

Figure: *Illustration of steps 1–4 for plotting a point on a Wulff net*

For an example of the use of the Wulff net, imagine that we have two copies of it on thin paper, one atop the other, aligned and tacked at their mutual centre. Suppose that we want to plot the point (0.321, 0.557, “0.766) on the lower unit hemisphere. This point lies on a line oriented 60° counterclockwise from the positive x-axis (or 30° clockwise from the positive y-axis) and 50° below the horizontal plane $z = 0$. Once these angles are known, there are four steps:

1. Using the grid lines, which are spaced 10° apart in the figures here, mark the point on the edge of the net that is 60° counterclockwise from the point (1, 0) (or 30° clockwise from the point (0, 1)).
2. Rotate the top net until this point is aligned with (1, 0) on the bottom net.
3. Using the grid lines on the bottom net, mark the point that is 50° towards the centre from that point.
4. Rotate the top net oppositely to how it was oriented before, to bring it back into alignment with the bottom net. The point marked in step 3 is then the projection that we wanted.

To plot other points, whose angles are not such round numbers as 60° and 50°, one must visually interpolate between the nearest grid lines. It is helpful to have a net with finer spacing than 10°; spacings of 2° are common.

To find the central angle between two points on the sphere based on their stereographic plot, overlay the plot on a Wulff net and rotate the plot about the centre until the two points lie on or near a meridian. Then measure the angle between them by counting grid lines along that meridian.

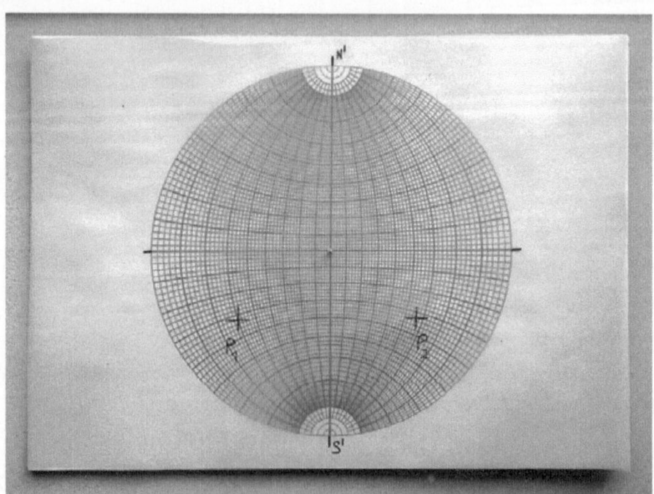

Figure: *Two points P_1 and P_2 are drawn on a transparent sheet tacked at the origin of a Wulff net.*

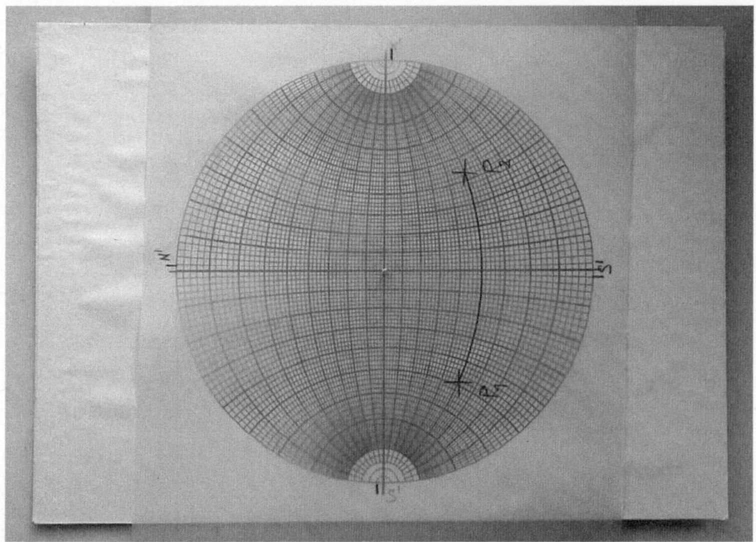

Figure: *The transparent sheet is rotated and the central angle is read along the common meridian to both points P_1 and P_2.*

Other Formulations and Generalizations

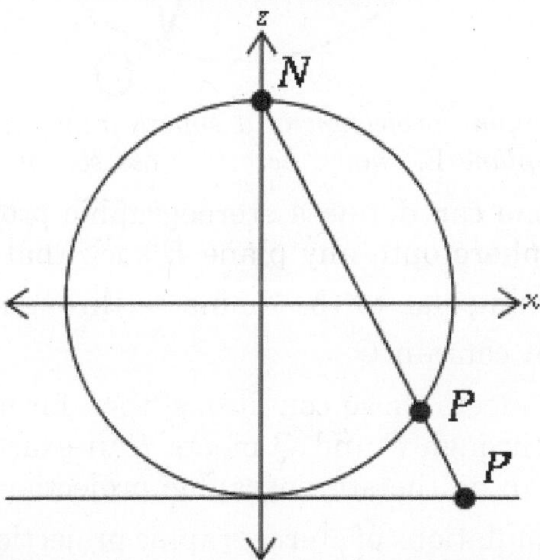

Figure: *Stereographic projection of the unit sphere from the north pole onto the plane z = –1, shown here in cross section*

Some authors define stereographic projection from the north pole (0, 0, 1) onto the plane $z = -1$, which is tangent to the unit sphere at the south pole (0, 0, –1). The values X and Y produced by this projection are exactly twice those produced by the equatorial projection described in the preceding section. For example, this projection sends the equator to the circle of radius 2 centred at the origin. While the equatorial projection produces no infinitesimal area distortion along the equator, this pole-tangent projection instead produces no infinitesimal area distortion at the south pole.

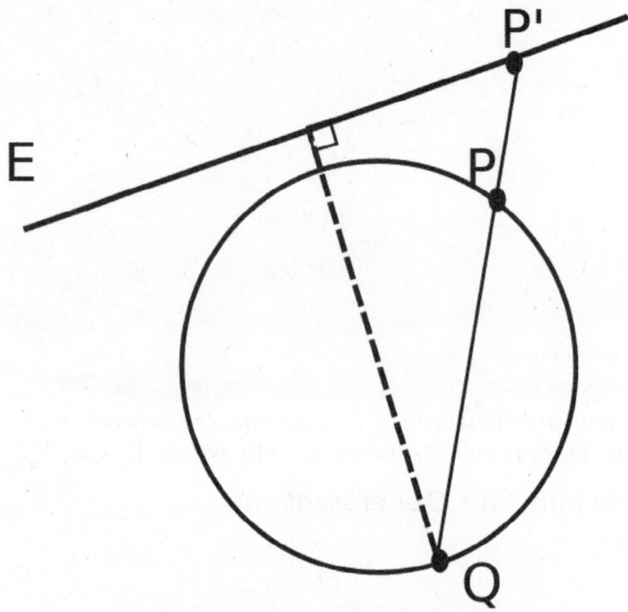

Figure: *Stereographic projection of a sphere from a point* Q *onto the plane* E, *shown here in cross section*

In general, one can define a stereographic projection from any point Q on the sphere onto any plane E such that

- E is perpendicular to the diametre through Q, and
- E does not contain Q.

As long as E meets these conditions, then for any point P other than Q the line through P and Q meets E in exactly one point P^2, which is defined to be the stereographic projection of P onto E.

All of the formulations of stereographic projection described thus far have the same essential properties. They are smooth bijections (diffeomorphisms) defined everywhere except at the projection point. They are conformal and not area-preserving.

More generally, stereographic projection may be applied to the n-sphere S^n in $(n + 1)$-dimensional Euclidean space E^{n+1}. If Q is a point of S^n and E a hyperplane in E^{n+1}, then the stereographic projection of a point P "S^n " $\{Q\}$ is the point P^2 of intersection of the line $\overline{QP}$ with E.

Still more generally, suppose that S is a (nonsingular) quadric hypersurface in the projective space P^{n+1}. By definition, S is the locus of zeros of a non-singular quadratic form $f(x_0, \ldots, x_{n+1})$ in the homogeneous coordinates x_i. Fix any point Q on S and a hyperplane E in P^{n+1} not containing Q.

Then the stereographic projection of a point P in $S - \{Q\}$ is the unique point of intersection of $\overline{QP}$ with E. As before, the stereographic projection is conformal and invertible outside of a "small" set. The stereographic projection presents the quadric hypersurface as a rational hypersurface. This construction plays a role in algebraic geometry and conformal geometry.

5

Applications within Mathematics

Complex Analysis

Although any stereographic projection misses one point on the sphere (the projection point), the entire sphere can be mapped using two projections from distinct projection points. In other words, the sphere can be covered by two stereographic parametrizations (the inverses of the projections) from the plane.

The parametrizations can be chosen to induce the same orientation on the sphere. Together, they describe the sphere as an oriented surface (or two-dimensional manifold).

This construction has special significance in complex analysis. The point (X, Y) in the real plane can be identified with the complex number $\zeta = X + iY$. The stereographic projection from the north pole onto the equatorial plane is then

$$\zeta = \frac{x+iy}{1-z},$$

$$(x,y,z) = \left(\frac{2\mathrm{Re}(\zeta)}{1+\bar{\zeta}\zeta}, \frac{2\mathrm{Im}(\zeta)}{1+\bar{\zeta}\zeta}, \frac{-1+\bar{\zeta}\zeta}{1+\bar{\zeta}\zeta}\right).$$

Similarly, letting $\partial = X - iY$ be another complex coordinate, the functions

$$\xi = \frac{x-iy}{1+z},$$

$$(x,y,z) = \left(\frac{2\mathrm{Re}(\xi)}{1+\bar{\xi}\xi}, \frac{2\mathrm{Im}(\xi)}{1+\bar{\xi}\xi}, \frac{1-\bar{\xi}\xi}{1+\bar{\xi}\xi}\right).$$

Define a stereographic projection from the south pole onto the equatorial plane. The transition maps between the ζ- and ∂-coordinates are then $\zeta = 1 / \zeta$ and $\zeta = 1 / \zeta$, with ζ approaching 0 as ζ goes to infinity, and *vice versa*. This facilitates an elegant and useful notion of infinity for the complex numbers and indeed an entire theory of meromorphic functions mapping to the Riemann sphere. The standard metric on the unit sphere agrees with the Fubini–Study metric on the Riemann sphere.

Visualization of Lines and Planes

The set of all lines through the origin in three-dimensional space forms a space called the real projective plane. This space is difficult to visualize, because it cannot be embedded in three-dimensional space. However, one can "almost" visualize it as a disk, as follows. Any line through the origin intersects the southern hemisphere $z \leq 0$ in a point, which can then be stereographically projected to a point on a disk.

Horizontal lines intersect the southern hemisphere in two antipodal points along the equator, either of which can be projected to the disk; it is understood that antipodal points on the boundary of the disk represent a single line. So any set of lines through the origin can be pictured, almost perfectly, as a set of points in a disk.

Also, every plane through the origin intersects the unit sphere in a great circle, called the *trace* of the plane. This circle maps to a circle under stereographic projection. So the projection lets us visualize planes as circular arcs in the disk. Prior to the availability of computers, stereographic projections with great circles often involved drawing large-radius arcs that required use of a beam compass. Computers now make this task much easier.

Further associated with each plane is a unique line, called the plane's *pole*, that passes through the origin and is perpendicular to the plane. This line can be plotted as a point on the disk just as any line through the origin can. So the stereographic projection also lets us visualize planes as points in the disk. For plots involving many planes, plotting their poles produces a less-cluttered picture than plotting their traces. This construction is used to visualize directional data in crystallography and geology, as described below.

Other Visualization

Stereographic projection is also applied to the visualization of polytopes. In a Schlegel diagram, an n-dimensional polytope in R^{n+1}

is projected onto an n-dimensional sphere, which is then stereographically projected onto R^n. The reduction from R^{n+1} to R^n can make the polytope easier to visualize and understand.

Arithmetic Geometry

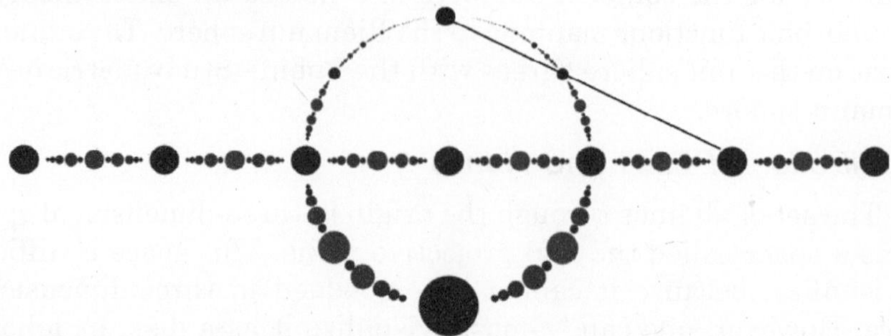

***Figure:** The rational points on a circle correspond, under stereographic projection, to the rational points of the line.*

In elementary arithmetic geometry, stereographic projection from the unit circle provides a means to describe all primitive Pythagorean triples. Specifically, stereographic projection from the north pole (0,1) onto the x-axis gives a one-to-one correspondence between the rational number points (x,y) on the unit circle (with $y \neq 1$) and the rational points of the x-axis. If $(m/n, 0)$ is a rational point on the x-axis, then its inverse stereographic projection is the point

$$\left(\frac{2mn}{n^2+m^2}, \frac{n^2-m^2}{n^2+m^2}\right)$$

which gives Euclid's formula for a Pythagorean triple.

Tangent half-angle Substitution

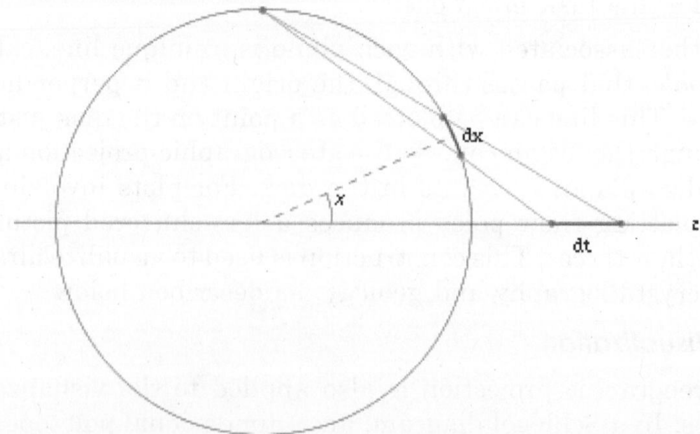

The pair of trigonometric functions (sin x, cos x) can be thought of as parametrizing the unit circle. The stereographic projection gives an alternative parametrization of the unit circle:

$$\cos x = \frac{t^2 - 1}{t^2 + 1}, \quad \sin x = \frac{2t}{t^2 + 1}.$$

Under this reparametrization, the length element dx of the unit circle goes over to

$$dx = \frac{2dt}{t^2 + 1}.$$

This substitution can sometimes simplify integrals involving trigonometric functions.

Applications to other Disciplines

Cartography

The fundamental problem of cartography is that no map from the sphere to the plane can accurately represent both angles (and thus shapes) and areas. In general, area-preserving map projections are preferred for statistical applications, because they behave well with respect to integration, while angle-preserving (conformal) map projections are preferred for navigation.

Stereographic projection falls into the second category. When the projection is centred at the Earth's north or south pole, it has additional desirable properties: It sends meridians to rays emanating from the origin and parallels to circles centred at the origin.

The stereographic is the only projection that maps all small circles to circles. This property is valuable in planetary mapping when craters are typical features.

Crystallography

In crystallography, the orientations of crystal axes and faces in three-dimensional space are a central geometric concern, for example in the interpretation of X-ray and electron diffraction patterns. These orientations can be visualized as in the section Visualization of lines and planes above. That is, crystal axes and poles to crystal planes are intersected with the northern hemisphere and then plotted using stereographic projection. A plot of poles is called a pole figure.

In electron diffraction, Kikuchi line pairs appear as bands decorating the intersection between lattice plane traces and the Ewald sphere thus providing *experimental access* to a crystal's stereographic projection. Model Kikuchi maps in reciprocal space, and fringe visibility maps for use with bend contours in direct space, thus act as road maps for exploring orientation space with crystals in the transmission electron microscope.

Anamorphosis

Anamorphosis is a distorted projection or perspective requiring the viewer to use special devices or occupy a specific vantage point to reconstitute the image. The word "anamorphosis" is derived from the Greek prefix ana-, meaning back or again, and the word morphe, meaning shape or form.

Types of Projection

There are two main types of anamorphosis: perspective (oblique) and mirror (catoptric). Examples of perspectival anamorphosis date to the early Renaissance (fifteenth century).

Examples of mirror anamorphosis were first created in the late Renaissance (sixteenth century). With mirror anamorphosis, a conical or cylindrical mirror is placed on the drawing or painting to transform a flat distorted image into a three-dimensional picture that can be viewed from many angles. The deformed image is painted on a plane surface surrounding the mirror. By looking uniquely into the mirror, the image appears undeformed. This process of anamorphosis made it possible to diffuse caricatures, erotic and scatological scenes and scenes of sorcery for a confidential public.

History

Leonardo's Eye (Leonardo da Vinci, c. 1485) is the earliest known definitive example of perspective anamorphosis in modern times. The prehistoric cave paintings at Lascaux may also possess this technique because the oblique angles of the cave would otherwise result in distorted figures from a viewer's perspective.

Hans Holbein the Younger is well known for incorporating this type of anamorphic trick. His painting *The Ambassadors* is the most famous example for anamorphosis, in which a distorted shape lies diagonally across the bottom of the frame. Viewing this from an acute angle transforms it into the plastic image of a skull. During the seventeenth century, Baroque trompe l'oeil murals often used this technique to combine actual architectural elements with an illusion. When standing in front of the art work in a specific spot, the architecture blends with the decorative painting. The dome and vault of the Church of St. Ignazio in Rome, painted by Andrea Pozzo, represented the pinnacle of illusion. Due to neighbouring monks complaining about blocked light, Pozzo was commissioned to paint the ceiling to look like the inside of a dome, instead of building a real dome. As the ceiling is flat, there is only one spot where the illusion is perfect and a dome looks real.

Anamorphosis could be used to conceal images for privacy or personal safety. A secret portrait of Bonnie Prince Charlie, held at the West Highland Museum, Scotland, is painted in a distorted manner on a tray and can only be recognised when a polished cylinder is placed in the correct position. To possess such an image would have been seen as treachery in the aftermath of the 1746 Battle of Culloden.

In the eighteenth and nineteenth centuries, anamorphic images had come to be used more as children's games than fine art. In the twentieth century some artists wanted to renew the technique of anamorphosis. Important to mention is Marcel Duchamp's interest in anamorphosis, some of his installations are paraphrases of anamorphoses. Salvador Dalí also used the effect in a number of his paintings. Jan Dibbets conceptual works, the so-called "perspective corrections" are examples of "linear" anamorphoses. In the late twentieth century, mirror anamorphosis was revived as children's toys and games.

"Anamorphic" Effects in the Work of Contemporary Artists

The Swedish artist Hans Hamngren produced and exhibited many examples of mirror anamorphosis in the 60s and 70s. Shigeo Fukuda, a Japanese artist, designed both types of anamorphosis in the 70s and 80s. Patrick Hughes (artist), Fujio Watanabe, William Kentridge, István Orosz, Felice Varini, Matthew Ngui, Kelly Houle, Nigel Williams, and Judy Grace are fine artists creating anamorphic images. Currently, Myrna Hoffman designs anamorphoses for children's interactive toys.

The Dutch artist Leon Keer made an optical illusion painting with a life-size mirror cylinder during the 3-D street painting event FringeMK in 2010.

Another form of anamorphic art is often called "Slant Art". Examples are the sidewalk chalk paintings of Kurt Wenner and Julian Beever where the chalk painting, the pavement and the architectural surroundings all become part of an illusion. Art of this style can be produced by taking a photograph of an object or setting at a sharp angle, then putting a grid over the photograph, another, elongated grid on the footpath based on a specific perspective, and reproducing exactly the contents of one into the other, one square at a time.

In November, 2011, an article published on the website of the Smithsonian magazine questioned if an anamorphic garden work by French artist François Abélenet featured in front of Paris' city hall, the Hôtel de Ville entitled, "Qui croire?" [Eng: "Who to believe?"] was the, "World's Greatest New Artwork?" citing an online video of it.

Practical Uses

Cinemascope, Panavision, Technirama and other widescreen formats use anamorphosis to project a wider image from a narrower film frame. The system of anamorphic projection can be seen quite commonly on text written at a very flat angle on roadways, such as "Bus Lane" or "Children Crossing", which is easily read by drivers who otherwise would have difficulty reading as the vehicle approaches the text; when the vehicle is nearly above the text, its true abnormally elongated shape can be seen. Similarly, in many sporting stadiums, especially in Rugby football in Australia, it is used to promote company brands which are painted onto the playing surface; from the television camera angle, the writing appear as signs standing vertically within the field of play.

On some 0.5 litre Sprite bottles in Europe, an extra "bar code" was present. When the bottle is tilted towards the mouth while drinking, the bar code resolves into writing due to the anamorphic effect.

Impossible Objects

In the twentieth century, artists began to play with perspective by drawing impossible objects. These objects included stairs that always go up or cubes where the back meets the front. Such works were popularized by artist M. C. Escher and mathematician Roger Penrose. Although referred to as impossible objects, such objects as the Necker Cube and the Penrose triangle can be built using anamorphosis. When viewed at a certain angle, such sculptures appear as the so-called impossible objects.

Popular Culture

Since 1993, Myrna Hoffman's company, OOZ & OZ, has been producing mirror anamorphosis art kits and activities for children. The image below is from Hoffman's "Morph-O-Scopes" Kit. Hoffman's kits have earned more than two dozen national toy awards. Rick Wakeman's 1976 album *No Earthly Connection* features front and back cover photographs that are mirror anamorphoses. The original vinyl release included a mirrored Mylar sheet which could be curled into a cylinder for viewing the images.

In the 2008 detective film *Anamorph*, the plot line revolves around the solution of gruesome anamorphically distorted images.

The 2009 Video Game Batman: Arkham Asylum has a series of riddles posed by the classic Batman antagonist The Riddler, the solution of which is based on perspective anamorphosis.

2.5D

2.5D ("two-and-a-half-dimensional"), 3/4 perspective and pseudo-3D are terms, mainly in the video game industry, used to describe either 2D graphical projections and similar techniques used to cause a series of images (or scenes) to simulate the appearance of being three-dimensional (3D) when in fact they are not, or gameplay in an otherwise three-dimensional video game that is restricted to a two-dimensional plane.

Common in video games, these projections have also been useful in geographic visualization (GVIS) to help understand visual-cognitive spatial representations or 3D visualization.

The terms 3/4 perspective and 3/4 view trace their origins to portraiture and facial recognition, where they are used to describe a view of a person's face which is partway between a frontal view and a side view.

Computer Graphics

Axonometric & Oblique Projection

In axonometric projection and oblique projection, two forms of parallel projection, the viewpoint is rotated slightly to reveal other facets of the environment than what are visible in a top-down perspective or side view, thereby producing a three-dimensional effect. An object is "considered to be in an inclined position resulting in foreshortening of all three axes", and the image is a "representation on a single plane (as a drawing surface) of a three-dimensional object placed at an angle to the plane of projection." Lines perpendicular to the plane become points, lines parallel to the plane have true length, and lines inclined to the plane are foreshortened.

They are popular camera perspectives among 2D video games, most commonly those released for 16-bit or earlier and handheld consoles, as well as in later strategy and role-playing video games. The advantage of these perspectives are that they combine the visibility and mobility of a top-down game with the character recognizability of a side-scrolling game. Thus the player can be presented an overview of the game world in the ability to see it from above, more or less, and with additional details in artwork made possible by using an angle: Instead of showing a humanoid in top-down perspective, as a head and shoulders seen from above, the entire body can be drawn when using a slanted angle; Turning a character around would reveal how it looks from the sides, the front and the back, while the top-down perspective will display the same head and shoulders regardless.

There are three main divisions of axonometric projection: *isometric* (equal measure), *dimetric* (symmetrical and unsymmetrical), and *trimetric* (single-view or only two sides). The most common of these drawing types in engineering drawing is isometric projection. This projection is tilted so that all three axes create equal angles at intervals of 120 degrees. The result is that all three axes are equally foreshortened. In video games, a form of dimetric projection with a 2:1 pixel ratio is more common due to the problems of anti-aliasing and square pixels found on most computer monitors.

In oblique projection typically all three axes are shown unforeshortened. All lines parallel to the axes are drawn to scale, and diagonals and curved lines are distorted. One tell-tale sign of oblique projection is that the face pointed towards the camera retains its right angles with respect to the image plane.

Two of the most consistent examples of oblique projection are the *The Legend of Zelda* series of games for the Game Boy, Game Boy Color, and Game Boy Advance; and the *Pokémon* series for the Nintendo DS. Examples of axonometric projection include the latter games in the *SimCity* series, and role-playing games such as *Diablo* and *Baldur's Gate*.

Billboarding

In three-dimensional scenes, the term billboarding is applied to a technique in which objects are sometimes represented by two-dimensional images applied to a single polygon which is typically kept perpendicular to the line of sight. The name refers to the fact that objects are seen as if drawn on a billboard. This is also known as a backdrop. This can be used to good effect for a significant performance boost when the geometry is sufficiently distant that it can be seamlessly replaced with a 2D sprite. In games, this technique is most frequently applied to objects such as particles (smoke, sparks, rain) and low-detail vegetation. A pioneer in the use of this technique was the game *Jurassic Park: Trespasser*. It has since become mainstream, and is found in many games such as *Rome: Total War*, where it is exploited to simultaneously display thousands of individual soldiers on a battlefield. Other examples include early first-person shooters like *Wolfenstein 3D*, *Doom*, *Hexen* and *Duke Nukem 3D*.

Skyboxes & Skydomes

Skyboxes and skydomes are methods used to easily create a background to make a game level look bigger than it really is. If the level is enclosed in a cube, the sky, distant mountains, distant buildings, and other unreachable objects are rendered onto the cube's faces using a technique called cube mapping, thus creating the illusion of distant three-dimensional surroundings. A *skydome* employs the same concept but uses a sphere or hemisphere instead of a cube.

As a viewer moves through a 3D scene, it is common for the skybox or skydome to remain stationary with respect to the viewer. This technique gives the skybox the illusion of being very far away since other objects in the scene appear to move, while the skybox does not. This imitates real life, where distant objects such as clouds, stars and even mountains appear to be stationary when the viewpoint is displaced by relatively small distances. Effectively, everything in a skybox will always appear to be infinitely distant from the viewer. This consequence of skyboxes dictates that designers should be careful not to carelessly include images of discrete objects in the textures of a skybox since the viewer may be able to perceive the inconsistencies of those objects' sizes as the scene is traversed.

Scaling Along the Z Axis

In some games, sprites are scaled larger or smaller depending on its distance to the player, producing the illusion of motion along the Z (forward) axis. Sega's 1986 video game *Out Run* is a good example of this technique.

In *Out Run*, the player drives a Ferrari into depth of the game window. The palms on the left and right side of the street are the same bitmap, but have been scaled to different sizes, creating the illusion that some are closer than others. The angles of movement are left and right and into the depth (while still capable of doing so technically, this game did not allow making a U-turn or going into reverse, therefore moving out of the depth, as this did not make sense to the high-speed game play and tense time limit). Notice the view is comparable to that which a driver would have in reality when driving a car. The position and size of any billboard is generated by a (complete 3D) perspective transformation as are the vertices of the poly-line representing the centre of the street.

Often the centre of the street is stored as a spline and sampled in a way that on straight streets every sampling point corresponds to one scan-line on the screen. Hills and curves lead to multiple points on one line and one has to be chosen. Or one line is without any point and has to be interpolated lineary from the adjacent lines. Very memory intensive billboards are used in *Out Run* to draw corn-fields and water waves which are wider than the screen even at the largest viewing distance and also in *Test Drive* to draw trees and cliffs.

Drakkhen was notable for being among the first role-playing video games to feature a three-dimensional playing field. However, it did not employ a conventional 3D game engine, instead emulating one using character-scaling algorithms.

The player's party travels overland on a flat terrain made up of vectors, on which 2D objects are zoomed. Drakkhen features an animated day-night cycle, and the ability to wander freely about the game world, both rarities for a game of its era. This type of engine was later used in the game *Eternam*.

When *Duke Nukem 3D* was converted to the Game.com console in 1996, the more conventional 2.5D game engine of the game was scrapped for a lower resource version. Thus unlike every other version of the game, Duke Nukem cannot turn; he can only move forwards, backwards, and strafe left or right whilst scaling the graphics larger and smaller. This implementation was first-person rather than third-person like *Drakkhen* and the game was marketed as having "3D-style" graphics.

Parallax Scrolling

Parallaxing refers to when a collection of 2D sprites or layers of sprites are made to move independently of each other and/or the background to create a sense of added depth. The technique grew out of the multiplane camera technique used in traditional animation since the 1940s. This type of graphical effect was first used in the 1982 arcade game *Moon Patrol*. Examples include the skies in *Rise of the Triad*, the arcade version of *Rygar*, *Sonic the Hedgehog*, *Street Fighter II* and *Dracula X Chronicles*.

Mode 7

Mode 7, a display system effect that included rotation and scaling, allowed for a 3D effect while moving in any direction without any actual 3D models, and was used to simulate 3D graphics on systems such as the SNES.

Bump, Normal and Parallax Mapping

Bump mapping, normal mapping and parallax mapping are techniques applied to textures in 3D rendering applications such as video games to simulate bumps and wrinkles on the surface of an object without using more polygons. To the end user, this means that textures such as stone walls will have more apparent depth and thus greater realism with less of an influence on the performance of the simulation.

Bump mapping is achieved by perturbing the surface normals of an object and using a grayscale image and the perturbed normal during illumination calculations. The result is an apparently bumpy surface rather than a perfectly smooth surface although the surface of the underlying object is not actually changed. Bump mapping was introduced by Blinn in 1978.

In normal mapping, the unit vector from the shading point to the light source is dotted with the unit vector normal to that surface, and the dot product is the intensity of the light on that surface. Imagine a polygonal model of a sphere - you can only approximate the shape of the surface. By using a 3-channel bitmapped image textured across the model, more detailed normal vector information can be encoded. Each channel in the bitmap corresponds to a spatial dimension (X, Y and Z). These spatial dimensions are relative to a constant coordinate system for object-space normal maps, or to a smoothly varying coordinate system (based on the derivatives of position with respect to texture coordinates) in the case of tangent-space normal maps. This adds much more detail to the surface of a model, especially in conjunction with advanced lighting techniques.

Parallax mapping (also called offset mapping or virtual displacement mapping) is an enhancement of the bump mapping and normal mapping techniques implemented by displacing the texture coordinates at a point on the rendered polygon by a function of the view angle in tangent space (the angle relative to the surface normal) and the value of the height map at that point. At steeper view-angles, the texture coordinates are displaced more, giving the illusion of depth due to parallax effects as the view changes.

Film and Animation Techniques

The term is also used to describe an animation effect commonly used in music videos and, more frequently, title sequences. Brought to wide attention by the motion picture *The Kid Stays in the Picture* based on the book by film producer Robert Evans, it involves the layering and animating of two-dimensional pictures in three dimensional space. Earlier examples of this technique include Liz Phair's music video Down directed by Rodney Ascher and "A Special Tree" directed by musician Giorgio Moroder.

Graphic Design

The term also refers to an often-used effect in the design of icons and graphical user interfaces (GUIs), where a slight 3D illusion is created by the presence of a virtual light source to the left (or in some cases right) side, and above a person's computer monitor. The light source itself is always invisible, but its effects are seen in the lighter colours for the top and left side, simulating reflection, and the darker colours to the right and below of such objects, simulating shadow.

An advanced version of this technique can be found in some specialised graphic design software, such as Pixologic's ZBrush. The idea is that the programme's canvas represents a normal 2D painting surface, but that the data structure that holds the pixel information is also able to store information with respect to a z-index, as well material settings, specularity, etc. Again, with this data it is thus possible to simulate lighting, shadows, and so forth.

3D Games With a two-dimensional Playing Field

The term "2.5D" is also applied (though mathematically incorrect) to 3D games that use polygonal graphics to render the world and/or characters, but whose gameplay is restricted to a 2D plane. Examples include *Pandemonium*, *Einhänder*, *Klonoa: Door to Phantomile*, *LittleBigPlanet*, *Nights into Dreams...* and *New Super Mario Bros. U*. The *Crash Bandicoot* series is sometimes referred to as 2.5D because the characters and scenery are rendered in 3D, yet most of its levels are not as free-roaming as "true" 3D platformers.

Some fighting games such as the *Super Smash Bros.* series, *Marvel Vs. Capcom 3, Street Fighter IV, Mortal Kombat, BlazBlue* and *Guilty Gear Xrd* also utilize 2.5D to showcase 3D backdrops and/or characters while limiting the action to a 2D plane.

In some games, such as *Goemon's Great Adventure* and *Pandemonium*, the area of gameplay can be described as a two-dimensional surface twisting and bending in a three-dimensional space. Inside this surface, the character and physics behave like in a traditional sidescrolling platformer. There are, however, a number of twists that aren't possible with normal sidescroller platformers: it is common in such games to let the two-dimensional plane cross itself or other planes on certain points, thus creating "track switches" in the course. Players can explore different areas of the 3D world that way or can be brought back to previous points seamlessy. Interactions with the "background" (non-accessible points in the 3D landscape) are also used extensively.

History

The first video games that used pseudo-3D were primarily arcade games, the earliest known examples dating back to the mid-1970s, when they began using microprocessors. In 1975, Taito released *Interceptor*, an early first-person shooter and combat flight simulator that involved piloting a jet fighter, using an eight-way joystick to aim with a crosshair and shoot at enemy aircraft that move in formations of two and increase/decrease in size depending on their distance to the player. In 1976, Sega released *Moto-Cross*, an early black-and-white motorbike racing video game, based on the motocross competition, that was most notable for introducing an early three-dimensional third-person perspective.

Later that year, Sega-Gremlin re-branded the game as *Fonz*, as a tie-in for the popular sitcom, *Happy Days*. Both versions of the game displayed a constantly changing forward-scrolling road and the player's bike in a third-person perspective where objects nearer to the player are larger than those nearer to the horizon, and the aim was to steer the vehicle across the road, racing against the clock, while avoiding any on-coming motorcycles or driving off the road. That same year also saw the release of two arcade games that extended the car driving subgenre into three dimensions with a first-person perspective: Sega's *Road Race*, which displayed a constantly changing forward-scrolling S-shaped road with two obstacle race cars moving along the road that the player must avoid crashing while racing against the clock, and Atari's *Night Driver*, which presented a series of posts by the edge of the road though there was no view of the road or the player's car. Games using vector graphics had an advantage in creating pseudo-3D effects. 1979's *Speed Freak* recreated the perspective of *Night Driver* in greater detail.

In 1979, Nintendo debuted *Radar Scope*, a shoot 'em up that introduced a three-dimensional third-person perspective to the genre, imitated years later by shooters such as Konami's *Juno First* and Activision's *Beamrider*. In 1980, Atari's *Battlezone* was a breakthrough for pseudo-3D gaming, recreating a 3D perspective with unprecedented realism, though the gameplay was still planar. It was followed up that same year by *Red Baron*, which used scaling vector images to create a forward scrolling rail shooter.

Sega's arcade shooter *Space Tactics*, released in 1980, allowed players to take aim using crosshairs and shoot lasers into the screen at enemies coming towards them, creating an early 3D effect. It was followed by other arcade shooters with a first-person perspective during the early 1980s, including Taito's 1981 release *Space Seeker*, and Sega's *Star Trek* in 1982. Sega's *SubRoc-3D* in 1982 also featured a first-person perspective and introduced the use of stereoscopic 3-D through a special eyepiece.

Sega's *Astron Belt* in 1983 was the first laserdisc video game, using full-motion video to display the graphics from a first-person perspective. Third-person rail shooters were also released in arcades at the time, including Sega's *Tac/Scan* in 1982, Nippon's *Ambush* in 1983, Nichibutsu's *Tube Panic* in 1983, and Sega's 1982 release *Buck Rogers: Planet of Zoom*, notable for its fast pseudo-3D scaling and detailed sprites.

In 1981, Sega's *Turbo* was the first racing game to feature a third-person perspective, rear view format. It was also the first racing game to use sprite scaling with full-colour graphics. *Pole Position* by Namco is one of the first racing games to use the trailing camera effect that is now so familiar. In this particular example, the effect was produced by linescroll—the practice of scrolling each line independently in order to warp an image. In this case, the warping would simulate curves and steering. To make the road appear to move towards the player, per-line color changes were used, though many console versions opted for palette animation instead.

Zaxxon, a shooter introduced by Sega in 1982, was the first game to use isometric axonometric projection, from which its name is derived. Though Zaxxon's playing field is semantically 3D, the game has many constraints which classify it as 2.5D: a fixed point of view, scene composition from sprites, and movements such as bullet shots restricted to straight lines along the axes. It was also one of the first video games to display shadows. The following year, Sega released the first pseudo-3D isometric platformer, *Congo Bongo*. Another early pseudo-3D platform game released that year was Konami's *Antarctic Adventure*, where the player controls a penguin in a forward-scrolling third-person perspective while having to jump over pits and obstacles. It was one of the earliest pseudo-3D games available on a computer, released for the MSX in 1983. That same year, Irem's *Moon Patrol* was a side-scrolling run & gun platform-shooter that introduced the use of layered parallax scrolling to give a pseudo-3D effect. In 1985, *Space Harrier* introduced Sega's "Super Scaler" technology that allowed pseudo-3D sprite-scaling at high frame rates, with the ability to scale 32,000 sprites and fill a moving landscape with them.

The first original home console game to use pseudo-3D, and also the first to use multiple camera angles mirrored on television sports broadcasts, was *Intellivision World Series Baseball* (1983) by Don Daglow and Eddie Dombrower, published by Mattel. Its television sports style of display was later adopted by 3D sports games and is now used by virtually all major team sports titles. In 1984, Sega ported several pseudo-3D arcade games to the Sega SG-1000 console, including a smooth conversion of the third-person pseudo-3D rail shooter *Buck Rogers: Planet of Zoom*.

With the advent of consoles and computer systems that were able to handle several thousands of polygons (the most basic element of *3D computer graphics*) per second and the usage of 3D specialized graphics processing unit, pseudo 3D became obsolete. But even today, there are computer systems in production, such as cellphones, which are often not powerful enough to display *true* 3D graphics, and therefore use pseudo-3D for that purpose. Interestingly, many games from the 1980s' *pseudo-3D arcade era* and *16-bit console era* are ported to these systems, giving the manufactures the possibility to earn revenues from games that are now nearly twenty years old.

By 1989, 2.5D representations were surfaces drawn with depth cues and a part of graphic libraries like GINO. 2.5D was also used in terrain modelling with software packages such as ISM from Dynamic

Graphics, GEOPAK from Uniras and the Intergraph DTM system. 2.5D surface techniques gained popularity within the geography community because of its ability to visualize the normal thickness to area ratio used in many geographic models; this ratio was very small and reflected the thinness of the object in relation to its width, which made it the object realistic in a specific plane. These representations were axiomatic in that the entire subsurface domain was not used or the entire domain could not be reconstructed; therefore, it used only a surface and a surface is one aspect not the full 3D identity.

The resurgence of 2.5D or visual analysis, in natural and earth science, has increased the role of computer systems in the creation of spatial information in mapping. GVIS has made real the search for unknowns, real-time interaction with spatial data, and control over map display and has paid particular attention to three-dimensional representations. Efforts in GVIS have attempted to expand higher dimensions and make them more visible; most efforts have focused on "tricking" vision into seeing three dimensions in a 2D plane. Much like 2.5D displays where the surface of a three dimensional object is represented but locations within the solid are distorted or not accessible.

Technical Aspects & Generalizations

The reason for using pseudo-3D instead of "real" 3D computer graphics is that the system that has to simulate a three dimensional looking graphic is not powerful enough to handle the calculation intensive routines of 3D computer graphics, yet is capable of using tricks of modifying 2D graphics like bitmaps. One of these tricks is to stretch a bitmap more and more, therefore making it larger with each step, as to give the effect of an object coming closer and closer towards the player.

Even simple shading and size of an image could be considered pseudo-3D, as shading makes it look more realistic. If the light in a 2D game were 2D, it would only be visible on the outline, and because outlines are often dark, they would not be very clearly visible. However, any visible shading would indicate the usage of pseudo-3D lighting and that the image uses pseudo-3D graphics. Changing the size of an image can cause the image to appear to be moving closer or further away, which could be considered simulating a third dimension.

Dimensions are the variables of the data and can be mapped to specific locations in space; 2D data can be given 3D volume by adding a value to the x, y, or z plane. "Assigning height to 2D regions of a topographic map" associating every 2D location with a height/elevation value creates a 2.5D projection; this is not considered a "true 3D representation", however is used like 3D visual representation to "simplify visual processing of imagery and the resulting spatial cognition".

Cutaway Drawing

A cutaway drawing, also called a *cutaway diagram* is a 3D graphics, drawing, diagram and or illustration, in which surface elements a three-dimensional model are selectively removed, to make internal features visible, but without sacrificing the outer context entirely.

Overview

According to Diepstraten et al. (2003) "the purpose of a cutaway drawing is to allow the viewer to have a look into an otherwise solid opaque object. Instead of letting the inner object shine through the surrounding surface, parts of outside object are simply removed. This produces a visual appearance as if someone had cutout a piece of the object or sliced it into parts. Cutaway illustrations avoid ambiguities with respect to spatial ordering, provide a sharp contrast between foreground and background objects, and facilitate a good understanding of spatial ordering".

Though cutaway drawing are not dimensioned manufacturing blueprints, they are meticulously drawn by a handful of devoted artists who either had access to manufacturing details or deduced them by observing the visible evidence of the hidden skeleton (e.g. rivet lines, etc.). The goal of this drawings in studies can be to identify common design patterns for particular vehicle classes. Thus, the accuracy of most of these drawings, while not 100 percent, is certainly high enough for this purpose.

The technique is used extensively in computer-aided design. It has also been incorporated into the user interface of some video games. In The Sims, for instance, users can select through a control panel whether to view the house they are building with no walls, cutaway walls, or full walls.

History

The cutaway view and the exploded view were minor graphic inventions of the Renaissance that also clarified pictorial representation. This cutaway view originates in the early fifteenth century notebooks of Marino Taccola (1382 – 1453). In the 16th century cutaway views in definite form were used in Georgius Agricola's (1494-1555) mining book *De Re Metallica* to illustrate underground operations. The 1556 book is a complete and systematic treatise on mining and extractive metallurgy, illustrated with many fine and interesting woodcuts which illustrate every conceivable process to extract ores from the ground and metal from the ore, and more besides. It shows the many watermills used in mining, such as the machine for lifting men and material into and out of a mine shaft.

The term "Cutaway drawing" was already in use in the 19th century but, became popular in the 1930s.

Technique

The location and shape to cut the outside object depends on many different factors, for example:

- the sizes and shapes of the inside and outside objects,
- the semantics of the objects,
- personal taste, etc.

These factors, according to Diepstraten et al. (2003), "can seldom be formalized in a simple algorithm, But the properties of cutaway can be distinguish in two classes of cutaways of a drawing":

- *cutout* : illustrations were the cutaway is retricted to very simple and regularly shaped of often only a small number of planar slices into the outside object.
- *breakaway* : a cutaway realized by a single hole in the outside of the object.

Fisheye Lens

A fisheye lens is an ultra wide-angle lens that produces strong visual distortion intended to create a wide panoramic or hemispherical image. Fisheye lenses achieve extremely wide angles of view by forgoing producing images with straight lines of perspective (rectilinear images), opting instead for a special mapping (for example: equisolid angle), which gives images a characteristic convex non-rectilinear appearance.

The term *fisheye* was coined in 1906 by American physicist and inventor Robert W. Wood based on how a fish would see an ultra-wide hemispherical view from beneath the water (a phenomenon known as Snell's window). Their first practical use was in the 1920s for use in meteorology to study cloud formation giving them the name "whole-sky lenses". The angle of view of a fisheye lens is usually between 100 and 180 degrees while the focal lengths depend on the film format they are designed for. Mass-produced fisheye lenses for photography first appeared in the early 1960s and are generally used for their unique, distorted appearance.

For the popular 35 mm film format, typical focal lengths of fisheye lenses are between 8 mm and 10 mm for circular images, and 15–16 mm for full-frame images.

For digital cameras using smaller electronic imagers such as 1/4" and 1/3" format CCD or CMOS sensors, the focal length of "miniature" fisheye lenses can be as short as 1 to 2mm.

These types of lenses also have other applications such as re-projecting images filmed through a fisheye lens, or created via computer generated graphics, onto hemispherical screens.

Fisheye lenses are also used for scientific photography such as recording of aurora and meteors, and to study plant canopy geometry and to calculate near-ground solar radiation.

They are also used as peephole door viewers to give the user a wide field of view.

Uses in Photography

In a circular fisheye lens, the image circle is *inscribed in* the film or sensor area; in a full-frame fisheye lens the image circle is *circumscribed around* the film or sensor area.

Further, different fisheye lenses distort images differently, and the manner of distortion is referred to as their mapping function. A common type for consumer use is *equisolid angle.*

Although there are digital fisheye effects available both in-camera and as computer software they can't extend the angle of view of the original images to the very large one of a true fisheye lens.

Circular Fisheye

The first types of fisheye lenses to be developed were "circular fisheye" — lenses which took in a 180° hemisphere and projected this as a circle within the film frame. Some circular fisheyes were available in orthographic projection models for scientific applications.

These have a 180° *vertical* angle of view, and the horizontal and diagonal angle of view are also 180°.

Most circular fisheye lenses cover a smaller image circle than rectilinear lenses, so the corners of the frame will be completely dark.

Full-frame Fisheye

Figure: *An example of full-frame fisheye used in a closed space (Nikkor 10.5mm)*

As fisheye lenses gained popularity in general photography, camera companies began manufacturing fisheye lenses that enlarged the image circle to cover the entire rectangular frame, called a "full-frame fisheye".

The picture angle produced by these lenses only measures 180 degrees when measured from corner to corner: these have a 180° *diagonal* angle of view, while the horizontal and vertical angles of view will be smaller; for an equisolid angle-type 15 mm full-frame fisheye, the horizontal FOV will be 147°, and the vertical FOV will be 94°. The first full-frame fisheye lens to be mass-produced was a 16 mm lens made by Nikon in the early 1970s. Digital cameras with APS-C sized sensors require a 10.5 mm lens (or, for Canon APS-C cameras, a 10 mm lens) to get the same effect as a 16 mm lens on a camera with full-frame sensor.

Focal Length

Sigma currently makes a 4.5mm fisheye lens that captures a 180 degree field of view on a crop body. Sunex also makes a 5.6mm fisheye lens that captures a circular 185 degree field of view on a 1.5x Nikon and 1.6x Canon DSLR cameras.

Nikon produced a 6 mm circular fisheye lens that was initially designed for an expedition to Antarctica. It featured a 220-degree field of view, designed to capture the entire sky and surrounding ground when pointed straight up. This lens is no longer manufactured by Nikon, and is used nowadays to produce interactive virtual-reality images such as QuickTime VR and IPIX. Because of its very wide field of view, it is very large and cumbersome — weighing 5.2 kilograms (11 lb) and having a diametre of 236 millimetres (9.3 in). It dwarfs a regular 35 mm SLR camera and has its own tripod mounting point, a feature normally seen in large long-focus or telephoto lenses to reduce strain on the lens mount because the lens is heavier than the camera. The lens is extremely rare.

An 8 mm fisheye lens, also made by Nikon, has proven useful for scientific purposes because of its equidistant (equiangular) projection, in which distance along the radius of the circular image is proportional to zenith angle.

Fisheye Lenses for APS-C Cameras

- Pentax DA 10-17mm lens (F3.5-4.5) / Tokina AF 10-17mm f/ 3.5-4.5 AT-X DX
- Samyang 8mm f3.5 fisheye. Notable for its stereographic projection.
- Sigma 10mm F2.8 EX DC Fisheye HSM lens
- Nikon AF DX Fisheye-Nikkor 10.5mm f/2.8G ED
- Sigma 4.5 mm *f*/2.8 EX DC Circular Fisheye HSM for APS-C sensors

Fisheye Lenses for 35 mm Cameras

Circular fisheye

- Peleng 8 mm *f*/3.5
- Sigma 8 mm *f*/4.0 EX DG
- Sigma 8 mm *f*/3.5 EX DG - replaces the Sigma 8 mm *f*/4.0 EX DG

Full-frame Fisheye

- Canon EF 15 mm *f*/2.8 (discontinued)
- Canon Fisheye FD 15mm *f*/2.8 (old lens, does not work on EF mount)
- Minolta AF 16mm f/2.8 Fisheye lens
- Sigma 15mm *f*/2.8 EX DG Diagonal Fisheye

Zoom Fisheye

Figure: *Canon 8–15mm at 8mm of BMW M3*

Canon EF 8–15mm f/4L Fisheye USM – lens can be used as both a full-frame fisheye and a circular fisheye on a 35 mm full-frame film or DSLR such as the canon 5d (Mark I – III) cameras, it can only be used as a cropped circular or as a full-frame fisheye on EOS DSLRs with APS-C/H size sensors (a zoom lock is included).

- Pentax SMC Pentax-F Fish-Eye 1:3.5–4.5 17–28mm – lens was born for full-frame film cameras, to take the place of the 16mm f/2.8 in the AF era. It starts from a 17mm full-frame fisheye and reaches the end of the excursion as an overdistorted 28mm. Was intended as a "special effect" lens and never had big sales.

Miniature Fisheye Lenses

Miniature digital cameras, especially when used as security cameras, often tend to have such lenses for similar reasons. Miniature fisheye lenses are designed for small-format CCD/CMOS imagers commonly used in consumer and security cameras.

Popular format sizes are 1/4" (active area 3.6mmx2.7mm), 1/3" (active area 4.8mmx3.6mm) and 1/2" (active area 6.6mmx4.8mm). Depending on the imager active area, the same lens can form a circular image on one imager (e.g. 1/2"), and a full frame on the other (e.g. 1/4").

Other Applications

- Many planetariums now use fisheye projection lenses to project the night sky or other digital content onto the interior of a dome.
- Flight simulators and visual combat simulators use fisheye projection lenses in order to create an immersive environment for pilots, air traffic controllers, or military personnel to train in.
- Similarly, the IMAX Dome (previously 'OMNIMAX') motion-picture format involves photography through a circular fisheye lens, and projection through the same onto a hemispherical screen.
- Scientists and resource managers (e.g., biologists, foresters, and meteorologists) use fisheye lenses for hemispherical photography to calculate plant canopy indices and near-ground solar radiation. Applications include evaluation of forest health, characterization of monarch butterfly winter roosting sites, and management of vineyards.
- Astronomers use fisheye lenses to capture cloud cover and light pollution data.
- Photographers and videographers use fisheye lenses so they can get the camera as close as possible for action shots whilst also capturing context, for example in skateboarding to focus on the board and still retain an image of the skater.
- The first music video to be shot completely with fisheye lens was for the Beastie Boys song "Shake Your Rump" in 1989.
- In Computer Graphics, circular fisheye images can be used to create environment map from physical world. One complete 180-degree wide angle fisheye image will fit to half of cubic mapping space if proper algorithm applies on. Environment maps can be used to render 3D object and virtual panoramic scene.

Mapping Function

The mapping of a sideways object leads to a picture position displacement from the image centre. The manner of this conversion is the mapping function. The distance of a point from the image centre 'r' is dependent on the focal length of the optical system 'f', and the angle from the optical axis 'θ', where 'θ' is in radians.

Normal (non-fisheye) Lens

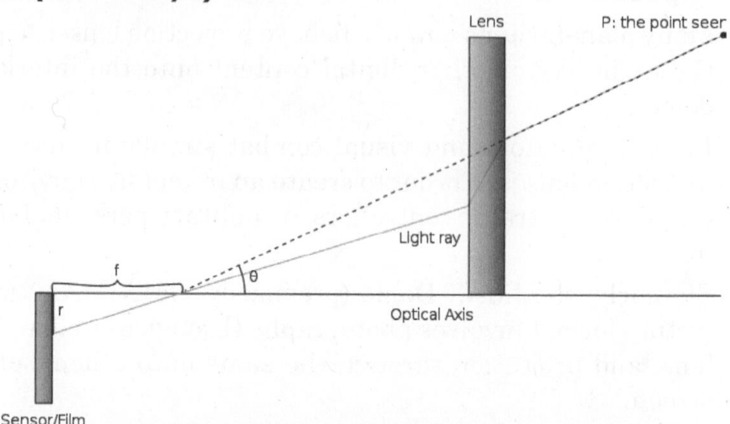

Figure: *Illustration of the conceptual significance of the* r, f, *and* θ *variables used in the mapping function specification.*

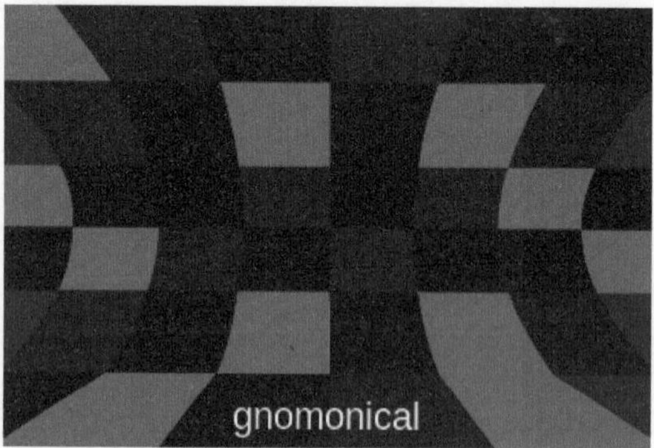

Figure: *Gnomonical*

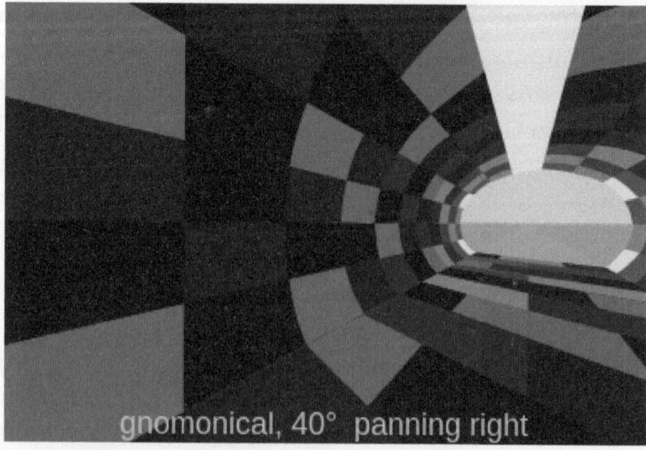

Figure: *Gnomonical, 40° right pan*

- Gnomonical (perspective): $r = f\tan(\theta)$. Works like the pinhole camera. Straight lines remain straight (distortion free). θ has to be smaller than 90°. The aperture angle is gaged symmetrically to the optical axis and has to be smaller than 180°. Large aperture angles are difficult to design and lead to high prices.

Fisheye Lens

Fisheye lenses can have many different mapping functions:

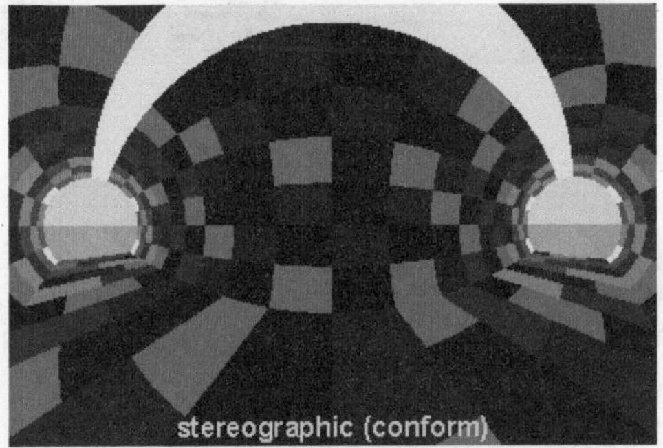

Figure: *Stereographic*

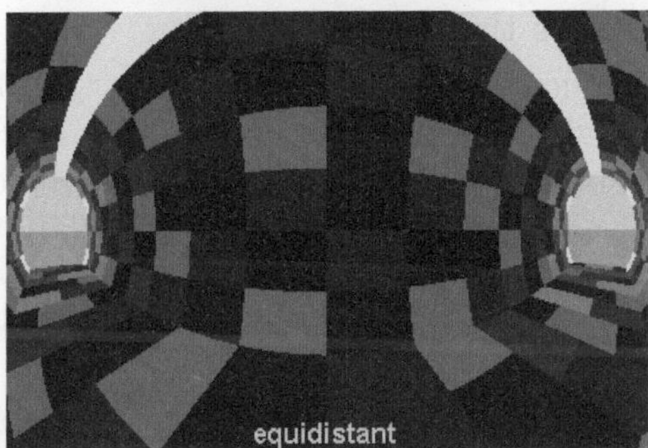

Figure: *Equidistant*

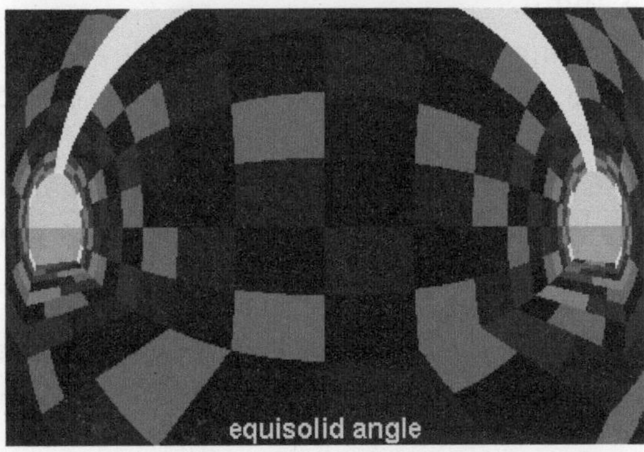

Figure: *Equisolid angle*

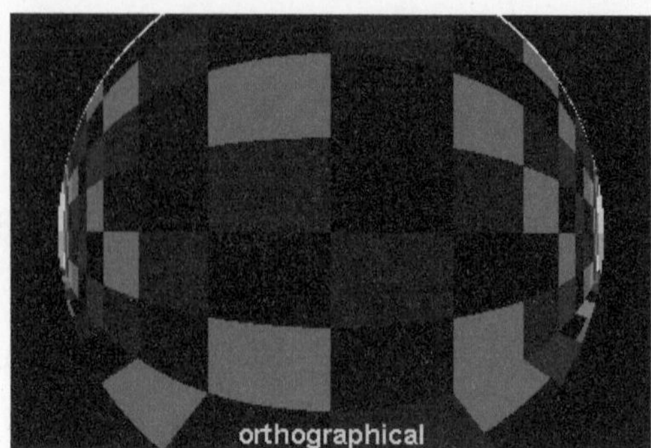

***Figure:** Orthographic*

- Stereographic (conform): $r = 2f\tan(\theta / 2)$. Maintains angles. This mapping would be ideal for photographers because it doesn't compress marginal objects as much. Samyang is the only manufacturer to produce this kind of fisheye lens, but it is available under different brand names. This mapping is easily implemented by software.
- Equidistant (linear scaled): $r = f\cdot\theta$. Maintains angular distances. Practical for angle measurement (e.g., star maps). PanoTools uses this type of mapping.
- Equisolid angle (equal area): $r = 2f\sin(\theta / 2)$. Maintains surface relations. Every pixel subtends an equal solid angle, or an equal area on the unit sphere. Looks like a mirror image on a ball, best special effect (unsophisticated distances), suitable for area comparison (clouds grade determination). This type is popular but it compresses marginal objects. The prices of these lenses are high, but not extreme.
- Orthographic: $r = f\sin(\theta)$. Maintains planar illuminance. Looks like an orb with the surroundings lying on < max. 180° aperture angle.
- Other mapping functions (for example Tailored Distortion lenses) are also possible for enhancing the off-axis resolution of fisheye lenses.

With appropriate software, the curvilinear images produced by a fisheye lens can be remapped to a conventional rectilinear projection. Although this entails some loss of detail at the edges of the frame, the technique can produce an image with a field of view greater than that of a conventional rectilinear lens. This is particularly useful for creating panoramic images. All types of fisheye lens bend straight lines. Aperture angles of 180° or more are possible only with large amounts of barrel distortion.

6

Electrical Drawing

An electrical drawing, is a type of technical drawing that shows information about power, lighting, and communication for an engineering or architectural project. Any electrical working drawing consists of "lines, symbols, dimensions, and notations to accurately convey an engineering's design to the workers, who install the electrical system on the job".

A complete set of working drawings for the average electrical system in large projects usually consists of:

- A plot plan showing the building's location and outside electrical wiring
- Floor plans showing the location of electrical systems on every floor
- Power-riser diagrams showing panel boards
- Control wiring diagrams
- Schedules and other information in combination with construction drawings.

Electrical drafters prepare wiring and layout diagrams used by workers who erect, install, and repair electrical equipment and wiring in communication centres, power plants, electrical distribution systems, and buildings.Mechanical systems drawing

Mechanical systems drawing is a type of technical drawing that shows information about heating, ventilating, and air conditioning. It is a powerful tool that helps analyze complex systems. These drawings are often a set of detailed drawings used for construction projects; it is a requirement for all HVAC work. They are based on the floor and reflected ceiling plans of the architect. After the mechanical drawings are complete, they become part of the construction drawings, which is then used to apply for a building permit. They are also used to determine the price of the project.

Sets of Drawings (United states)

General Arrangement Drawing

The general arrangement shows all necessary information: overall dimensions, installation details, overall weight/mass, weight of sub systems, and service supply details.

Arrangement Drawing

Arrangement drawings include information about the self-contained units that make up the system: table of parts, fabrication and detail drawing, overall dimension, weight/mass, lifting points, and information needed to construct, test, lift, transport, and install the equipment.

These drawings should show at least three different orthographic views and clear details of all the components and how they are assembled.

Assembly Drawing

The assembly drawing typically includes three orthographic views of the system: overall dimensions, weight and mass, identification of all the components, quantities of material, supply details, list of reference drawings, and notes. Assembly drawings detail how certain component parts are assembled.

Detail Drawing

In detail drawings, components used to build the mechanical system are described in some detail to show that the designer's specifications are met: relevant codes, standards, geometry, weight, mass, material, heat treatment requirements, surface texture, size tolerances, and geometric tolerances.

Fabrication Drawings

A fabrication is made up of many different parts. A fabrication drawing has a list of parts that make up the fabrication. In the list, parts are identified (balloons and leader lines) and complex details are included: welding details, material standards, codes, and tolerances, and details about heat/stress treatments.

Tender Drawings

Drawings produced to enable those tendering to interpret the design and to submit a tender for executing all or any part of the Works as defined elsewhere.

Sketch Drawing

Line diagrams and layouts indicating basic proposals, location of main items of plant, routes of main pipes, air ducts and cable runs in such detail as to illustrate the incorporation of the engineering services within the project as a whole.

Schematic Drawing

The schematic is a line diagram, not necessarily to scale, that describes interconnection of components in a system. The main features of a schematic drawing show:

- A two dimensional layout with divisions that show distribution of the system between building levels, or an isometric-style layout that shows distribution of systems across individual floor levels
- All functional components that make up the system, i.e., plant items, pumps, fans, valves, strainers, terminals, electrical switchgear, distribution and components
- Symbols and line conventions, in accordance with industry standard guidance
- Labels for pipe, duct, and cable sizes where not shown elsewhere
- Components that have a sensing and control function, and links between them—building management systems, fire alarms and HV controls
- Major components, so their whereabouts in specifications and other drawings can be easily determined

Detailed Design Drawing

A drawing showing the intended locations of plant items and service routes in such detail as to indicate the design intent. The main features of detailed design drawings should be as follows:

- Plan layouts to a scale of at least 1:100.
- Plant areas to a scale of at least 1:50 and accompanied by cross-sections.
- The drawing don't indicate precise positions of services, but should be feasible to install the services within the general

routes indicated. It should be possible to produce co-ordination drawings or installation drawings without major re-routing of the services.

- Represent pipework by single line layouts.
- Represent ductwork by either double or single line layouts as required to ensure that the routes indicated are feasible.
- Indicate on the drawing the space available for major service routing in both horizontal and vertical planes.

Installation Drawing

A drawing based on the detailed drawing or co-ordination drawing with the primary purpose of defining that information needed by the tradesmen on site to install the works. The main features of typical installation drawings are:

- Plan layouts to a scale of at least 1:50, accompanied by cross-sections to a scale of at least 1:20 for all congested areas
- A spatially coordinated drawing, i.e., show no physical location clashes between the system components
- Allowance for inclusion of all supports and fixtures necessary to install the works
- Allowance for the service at its widest point for spaces between pipe and duct runs, for insulation, standard fitting dimensions, and joint widths
- Installation details provided from shop drawings
- Installation working space; space to facilitate commissioning and space to allow on-going operation and maintenance in accordance with the relevant health and safety requirements
- Plant and equipment including alternatives and options
- Dimensions where services positioning is important enough not to installers
- Plant room layouts to a scale of at least 1:20, accompanied by cross-sections and elevations to a scale of at least 1:20

Record Drawing, or As Installed Drawing

A drawing showing the building and services installations as installed at the date of practical completion. Generally the record drawing is a development of the installation drawing. The main features of the record drawings should be as follows.

- Provide a record of the locations of all the systems and components installed including pumps, fans, valves, strainers, terminals, electrical switchgear, distribution and components.
- Use a scale not less than that of the installation drawings.
- Have marked on the drawings the positions of access points for operating and maintenance purposes.
- The drawings should not be dimensioned unless the inclusion of a dimension is considered necessary for location.

Builder's work Drawing

Design stage

These drawings show the provisions required to accommodate the services that significantly affect the design of the building structure, fabric, and external works. This includes drawings (and schedules) of work the building trade carries out, or that must be cost-estimated at the design stage, e.g., plant bases

Installation Stage

These drawings show requirements for building works necessary to facilitate installing the engineering services (other than where it is appropriate to mark out on site). Information on these drawing includes details of all:

- Bases for plant formed in concrete, brickwork or blockwork, to a scale of not less than 1:20
- Attendant builders work, holes, chases, etc. for conduits, cables and trunking etc. and any item where access for a function of the installation is required to a scale of not less than 1:100
- Purpose made brackets for supporting service or plant/ equipment to a scale of not less than 1:50
- Accesses into ceilings, ducts, etc. at a scale of not less than 1:50
- Special fixings, inserts, brackets, anchors, suspensions, supports etc. at a scale of not less than 1:20
- Sleeves, puddle flanges, access chambers at a scale not less than 1:20

Details to Include

- Size, type, and layout of ducting
- Diffusers, heat registers, return air grilles, dampers

- Turning vanes, ductwork insulation
- HVAC unit
- Thermostats
- Electrical, water, and/or gas connections
- Ventilation
- Exhaust fans
- Symbol legend, general notes and specific key notes
- Heating and/or cooling load summary
- Connection to existing systems
- Demolition of part or all of existing systems
- Smoke detector and firestat re-ducting
- Thermostat programming
- Heat loss and heat gain calculations
- Special condition

Job Outlook

About 80,000 jobs are held by mechanical drafters in the United States of America during 2008. From 2008 to 2018, mechanical drafting hiring rate is expected to neither increase nor decrease.

It is encouraged to either take two additional years of training in drafting school after high school or attend a four-year college/ university to develop better technical skills and gain more experience with CADD (computer-aided design).

Income of Mechanical Drafters in 2008

- Lowest 10% made $29,390.
- Highest 10% made $71,340.
- Middle 50% made between $36,490 to $59,010.
- Median: $46,640.

ADDA-Certification

The American Design Drafting Association (ADDA) has developed a Drafter Certification Test. The test assesses the drafter's skill in basic drafting concepts:

Geometric construction, working drawings, and architectural terms and standards. The test is administered periodically at ADDA-authorized sites.

Regulations in Canada

Mechanical system drawings must abide by all of the following regulations: the National Building Code of Canada, the National Fire Code, and Model National Energy Code of Canada for Buildings. For residential projects, The National Housing Code of Canada and the Model National Energy Code of Canada for Houses must also be followed. These drawings must also adhere to local and provincial codes and bylaws.

Drafting Machine

***Figure:** Drafting machine attached to a drawing board.*

A drafting machine is a very useful tool in technical drawing, consisting of a pair of scales mounted to form a right angle on an articulated protractor head that allows an angular rotation.

The protractor head (two scales and protractor mechanism) is able to move freely across the surface of the drawing board, sliding on two guides directly or indirectly anchored to the drawing board. These guides, which act separately, ensure the movement of the set in the horizontal or vertical direction of the drawing board, and can be locked independently of each other.

Drafting machines were present in the design offices of European companies since the 1920s. It is curious to see how the Encyclopædia Britannica explicitly specifies 1930 as the year this tool was introduced, but an advertisement of "Memorie di architettura pratica" from 1913 places it twenty years before this date — at least in Italy.

In the older design sets, the movement of the protractor head was assured by a pantograph system that could keep the head in the same angular position throughout its range of motion. The arms were balanced by a system of counterweights or springs.

Typically, the machine is mounted on a drawing board with a hard and smooth surface, anchored to a base that allows its tilting and lifting. Thus, the realization of a drawing can be achieved in the most convenient way on a working surface that can be tilted at any angle from horizontal to vertical.

There are special versions for A0 double-sized boards, to make large drawings, or copying-boards with background illumination, which have all that is necessary to provide specific support.

With the drafting machine one can perform a series of drawing operations that otherwise could only be achieved with a much more complex use of the classic ruler square and protractor, as, for example, drawing parallel lines, orthogonal lines, inclined lines according to a preset angle, measurement of angles, etc.

With the development of computer-aided design (CAD), the use of drafting machines, especially in the professional sector, has been drastically reduced, and nowadays the role for achieving drawings on paper is relegated to the computer-driven plotter.

Drawing Board

A drawing board (also drawing table, drafting table or architect's table) is, in its antique form, a kind of multipurpose desk which can be used for any kind of drawing, writing or impromptu sketching on a large sheet of paper or for reading a large format book or other oversized document or for drafting precise technical illustrations. The drawing table used to be a frequent companion to a pedestal desk in a gentleman's study or private library, during the preindustrial and early industrial era.

During the Industrial Revolution draftsmanship gradually became a specialized trade and drawing tables slowly moved out of the libraries and offices of most gentlemen. They became more utilitarian and were built of steel and plastic instead of fine woods and brass. More recently engineers and draftsmen use the drawing board for making and modifying drawings on paper with ink or pencil. Different drawing instruments (set square, protractor, etc.) are used on it to draw parallel, perpendicular or oblique lines. There are instruments for drawing circles, arcs, other curves and symbols too (compass, French curve, stencil, etc.). However, with the gradual introduction of computer aided drafting and design (CADD or CAD) in the last decades of the 20th century and the first of the 21st century, the drawing board is becoming less common.

A drawing table is also sometimes called a mechanical desk because, for several centuries, most mechanical desks were drawing tables. Unlike the gadgety mechanical desks of the second part of the 18th century, however, the mechanical parts of drawing tables were usually limited to notches, ratchets, and perhaps a few simple gears, or levers or cogs to elevate and incline the working surface.

Very often a drawing table could look like a writing table or even a pedestal desk when the working surface was set at the horizontal and the height adjusted to 29 inches, in order to use it as a "normal" desk. The only giveaway was usually a lip on one of the sides of the desktop. This lip or edge stopped paper or books from sliding when the surface was given an angle. It was also sometimes used to hold writing implements. When the working surface was extended at its full height, a drawing table could be used as a standing desk.

Many reproductions have been made and are still being produced of drawing tables, copying the period styles they were originally made in during the 18th and 19th centuries.

Modern-day Idiom

The expression *"back to the drawing board"* is used when a plan or course of action needs to be changed, often drastically; usually due a very unsuccessful result; e.g., "The battle plan, the result of months of conferences, failed because the enemy retreated too far back. It was back to the drawing board for the army captains."

The phrase was coined in the caption to a Peter Arno cartoon of The New Yorker of March 1, 1941 (cartoon)

Contemporary Drafting Tables

Despite the prevalence of computer aided drafting, many older architects and even some structural designers still rely on paper and pencil graphics produced on a drafting table. Modern drafting tables typically rely on a steel frame. Steel provides as much strength as the old oak drafting table frames and much easier portability. Typically the drafting board surface is a thick sheet of compressed fibreboard with sheets of formica laminated to all its surfaces. The drafting board surface is usually secured to the frame by screws which can easily be removed for drafting table transportation.

The steel frame allows mechanical linkages to be installed that control both the height and angle of the drafting board surface. Typically, a single foot pedal is used to control a clutch which clamps the board in the desired position. A heavy counterweight full of lead shot is installed in the steel linkage so that if the pedal is accidentally released, the drafting board will not spring into the upright position and injure the user. Drafting table linkages and clutches have to be maintained to ensure that this safety mechanism counterbalances the weight of the table surface.

The drafting table surface is usually covered with a thin vinyl sheet called a board cover. This provides an optimum surface for pen and pencil drafting. It allows compasses and dividers to be used without damaging the wooden surface of the board. A board cover must be frequently cleaned to prevent graphite buildup from making new drawings dirty. At the bottom edge of the table, a single strip of aluminum or steel may serve as a place to rest drafting pencils. More purpose-built trays are also used which hold pencils even while the board is being adjusted.

Various types of drafting machine may be attached to the board surface to assist the draftsperson or artist. Parallel rules often span the entire width of the board and are so named because they remain parallel to the top edge of the board as they are moved up and down. Drafting machines use precalibrated scales and built in protractors to allow accurate drawing measurement.

Some drafting tables incorporate electric motors to provide the up and down and angle adjustment of the drafting table surface. These tables are at least as heavy as the original oak and brass drafting tables and so sacrifice portability for the convenience of pushbutton table adjustment.

Drafter

Drafter, draughtsman (BrE) or draftsman (AmE), can mean two things:

1. A visual artist who specializes in artistic drawings.
2. A person who prepares technical drawings and plans under the direction of an architect or engineer.

Overview

In the past, drafters sat at drawing boards and used pencils, pens, compasses, protractors, triangles, and other drafting devices to prepare a drawing by hand. During the late 1980s and early 1990s, drawing boards became largely superseded by the application of Computer Aided Design and Drafting (CADD) systems for the creation of technical drawings. Consequently, some drafters may be referred to as CADD operators.

With CADD or AutoCad systems, drafters can create and store drawings electronically so that they can be viewed, printed, or programmed directly into automated manufacturing systems. CADD systems also permit drafters to quickly prepare variations of a design. Although drafters use CADD extensively, it is only a tool. Drafters still need knowledge of traditional drafting techniques, in addition to CADD skills. Despite the nearly universal use of CADD systems, manual drafting and sketching are used in certain applications.

Drafters' drawings provide visual guidelines and show how to construct a product or structure. Drawings include technical details and specify dimensions, materials, and procedures. Drafters fill in technical details using drawings, rough sketches, specifications, and calculations made by engineers, surveyors, architects, or scientists. For example, drafters use their knowledge of standardized building techniques to draw in the details of a structure. Some use their understanding of engineering and manufacturing theory and standards to draw the parts of a machine; they determine design elements, such as the numbers and kinds of fasteners needed to assemble the machine. Drafters use technical handbooks, tables, calculators, and computers to complete their work.

Specialties

Drafting work has many specialties:

- Aeronautical drafters prepare engineering drawings detailing plans and specifications used in the manufacture of aircraft, missiles, and related parts.
- Architectural drafters draw architectural and structural features of buildings and other structures. These workers may

specialize in a type of structure, such as residential or commercial, or in a kind of material used, such as reinforced concrete, masonry, steel, or timber.

- Civil drafters prepare drawings and topographical and relief maps used in major construction or civil engineering projects such as highways, bridges, pipelines, flood control projects, and water and sewage systems.
- Electrical drafters prepare wiring and layout diagrams used by workers who erect, install, and repair electrical equipment and wiring in communication centres, power plants, electrical distribution systems, and buildings.
- Electronics drafters draw wiring diagrams, circuit board assembly diagrams, schematics, and layout drawings used in the manufacture, installation, and repair of electronic devices and components.
- Mechanical drafters prepare drawings showing the detail and assembly of a wide variety of machinery and mechanical devices, indicating dimensions, fastening methods, and other requirements.
- Process piping or pipeline drafters prepare drawings used in the layout, construction, and operation of oil and gas fields, refineries, chemical plants, and process piping systems.
- Photovoltaic drafters prepare drawings showing inverter Pad location drawings and slab construction drawings, also prepare specific photovoltaic system assembly details and some wiring diagrams.

Work Environment

Drafters usually work in offices, seated at adjustable drawing boards or drafting tables when doing manual drawings, although most drafters work at computer terminals much of the time. Because they spend long periods in front of computers doing detailed work, drafters may be susceptible to eyestrain, back discomfort, and hand and wrist problems. Most drafters work standard 40-hour weeks; only a small number work part-time.

Education and Training

High school courses in mathematics, science, computer technology, design, computer graphics, and, where available, drafting are useful for people considering a drafting career. Employers prefer applicants who have also completed training after high school at a technical institute, community college, or four-year college or university. The kind and quality of drafting training programmes vary considerably so prospective students should be careful in selecting a programme. They should contact prospective employers to ask which schools they prefer and contact schools to ask for information about the kinds of jobs their graduates have, the type and condition of instructional facilities and equipment, and teacher qualifications.

Technical institutes offer intensive technical training, but they provide a less general education than do community colleges. Either certificates or diplomas may be awarded. Many technical institutes offer two-year associate degree programmes, which are similar to, or part of, the programmes offered by community colleges or state university systems. Their programmes vary considerably in length and in the type of courses offered. Some public vocational-technical schools serve local students and emphasize the type of training preferred by local employers. Most require a high school diploma or its equivalent for admission. Other technical institutes are run by private, often for-profit, organizations sometimes called proprietary schools.

Flat spline

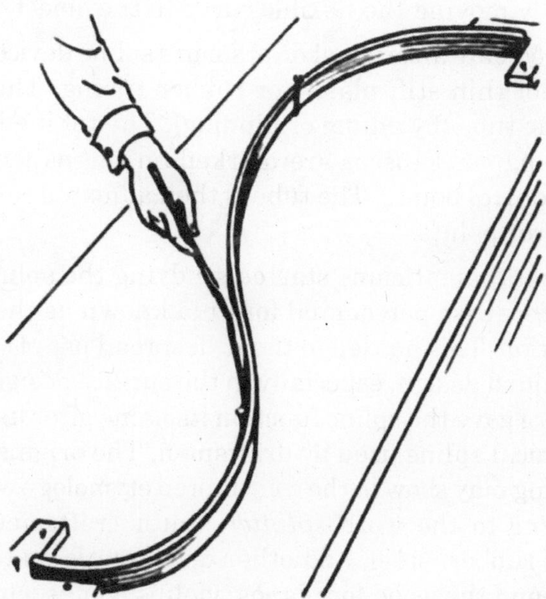

Figure: *A spline*

A spline or the more modern term flexible curve consists of a long strip fixed in position at a number of points that relaxes to form and hold a smooth curve passing through those points for the purpose of transferring that curve to another material.

Before computers were used for creating engineering designs, drafting tools were employed by designers drawing by hand. To draw curves, especially for shipbuilding, draftsmen often used long, thin, flexible strips of wood, plastic, or metal called splines (or laths, not to be confused with lathes). The splines were held in place with lead weights (called ducks because of their duck-like shape). The elasticity of the spline material combined with the constraint of the control points, or knots, would cause the strip to take the shape that minimized the energy required for bending it between the fixed points, this being the smoothest possible shape.

Splines are more recently referred to as flexible curves and perform much of the original function. The main difference between splines and flexible curves is that the control points of flexible curves are entirely internal in their housing. This has one advantage over splines: whereas the draftsman had to first set up the spline and control points before moving the object to be marked to the spline in that order only, with flexible curves the draftsman can also set up the flexible curve before carefully moving the flexible curve to the object to be marked.

One can recreate an original draftsman's spline device with weights and a length of thin stiff plastic or rubber tubing. The weights are attached to the tube (by gluing or pinning). The tubing is then placed over drawing paper. Crosses are marked on the paper to designate the knots or control points. The tube is then adjusted so that it passes over the control points.

In 1946, mathematicians started studying the spline shape, and derived the piecewise polynomial formula known as the spline curve or spline function. This has led to the widespread use of such functions in computer-aided design, especially in the surface designs of vehicles. I. J. Schoenberg gave the spline function its name after its resemblance to the mechanical spline used by draftsmen. The origins of the spline in wood-working may show in the conjectured etymology which connects the word *spline* to the word *splinter*. Later craftsmen have made splines out of rubber, steel, and other elastomeric materials. Spline devices help bend the wood for pianos, violins, violas, etc. The Wright brothers used one to shape the wings of their aircraft.

Floor Plan

In architecture and building engineering, a floor plan otherwise known as a scottish plan is a drawing to scale, showing a view from above, of the relationships between rooms, spaces and other physical features at one level of a structure.

Dimensions are usually drawn between the walls to specify room sizes and wall lengths. Floor plans may also include details of fixtures like sinks, water heaters, furnaces, etc. Floor plans may include notes for construction to specify finishes, construction methods, or symbols for electrical items.

It is also called a "plan" which is a measured plane typically projected at the floor height of 4', as opposed to an "elevation" which is a measured plane projected from the side of a building, along its height, or a section or "cross section" where a building, is cut along an axis to reveal the interior structure.

Overview

Similar to a map the orientation of the view is downward from above, but unlike a conventional map, a plan is understood to be drawn at a particular vertical position (commonly at about 4 feet above the floor). Objects below this level are seen, objects at this level are shown 'cut' in plan-section, and objects above this vertical position within the structure are omitted or shown dashed. Plan view or "planform" is defined as a vertical orthographic projection of an object on a horizontal plane, like a map.

The term may be used in general to describe any drawing showing the physical layout of objects.

For example, it may denote the arrangement of the displayed objects at an exhibition, or the arrangement of exhibitor booths at a convention. Now drawings are reproduced using plotters and large format xerographic copiers.

A "reflected ceiling plan" shows a view of the room as if looking from above, through the ceiling, at a mirror installed one foot below the ceiling level, which shows the *reflected* image of the ceiling above. This convention maintains the same orientation of the floor and ceilings plans - looking down from above. Reflected Ceiling Plans or RCP's are used by designers and architects to demonstrate lighting, visible mechanical features, and ceiling forms as part of the documents provided for construction.

Floor Plan Topics

Building Blocks

A floor plan is not a top view or birds eye view. It is a measured drawing to scale of the layout of a floor in a building. A "top view" or "bird's eye view" does not show an orthogonally projected plane cut at the typical 4' height above the floor level. A floor plan could show:

- Interior walls and hallways
- Restrooms
- Windows and doors
- Appliances such as stoves, refrigerators, water heater etc.
- Interior features such as fireplaces, saunas and whirlpools
- The use of all rooms shall be indicated

Plan View

A plan view is an orthographic projection of a 3-dimensional object from the position of a horizontal plane through the object. In other words, a plan is a section viewed from the top. In such views, the portion of the object in above the plane is omitted to reveal what lies beyond. In the case of a floor plan, the roof and upper portion of the walls may be omitted.

Roof plans are orthographic projections, but they are not sections as their viewing plane is outside of the object.

A plan is a common method of depicting the internal arrangement of a 3-dimensional object in two dimensions. It is often used in technical drawing and is traditionally crosshatched. The style of crosshatching indicates the type of material the section passes through.

French Curve

A French curve is a template made out of metal, wood or plastic composed of many different curves. It is used in manual drafting to draw smooth curves of varying radii. The shapes are segments of the Euler spiral or clothoid curve. The curve is placed on the drawing material, and a pencil, knife or other implement is traced around its curves to produce the desired result.

Modern Successors

Modern computer-aided drafting (CAD) systems use vector-based graphics to achieve a precise radius, so no template is required. Digital computers can also be used to generate a set of coordinates

that accurately describe an arbitrary curve, and the points can be connected with line segments to approximate the curve with a high degree of accuracy. Some computer graphic systems make use of Bézier splines, which allow a curve to be bent in real time on a display screen to follow a set of coordinates, much in the way a French curve would be placed on a set of three or four points on paper.

Graph Paper

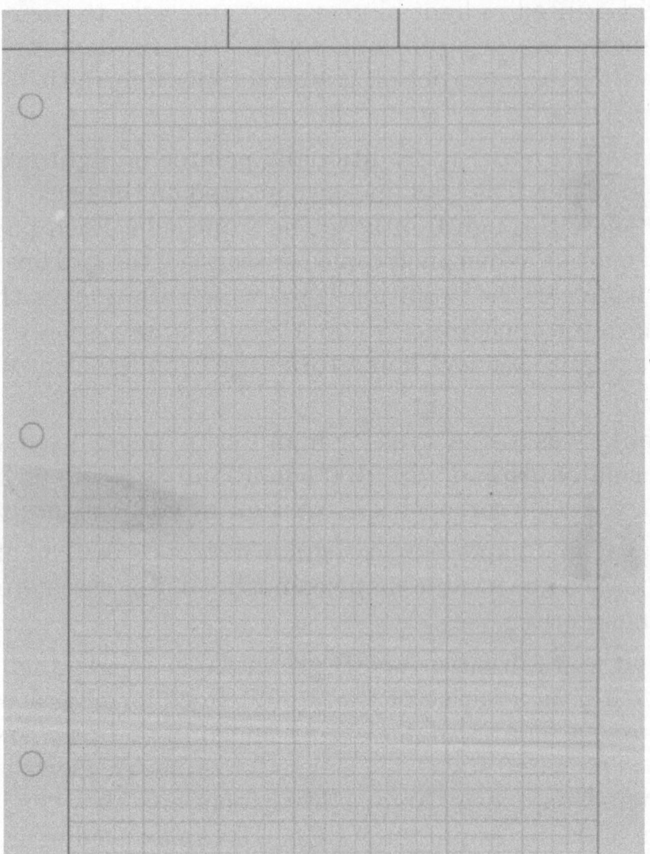

Figure: *Engineering paper*

Graph paper, graphing paper, or millimetre paper is writing paper that is printed with fine lines making up a regular grid. The lines are often used as guides for plotting mathematical functions or experimental data and drawing diagrams. It is commonly found in mathematics and engineering education settings and in laboratory notebooks. Graph paper is available either as loose leaf paper or bound in notebooks. It is becoming less common as computer software

such as spreadsheets and plotting software has supplanted many of the former uses of graph paper. Some users of graph paper now print pdf images of the grid pattern as needed rather than buying it pre-printed. There are websites that will generate pdfs of graph paper to the user's specifications.

Formats

- Quad paper is a common form of graph paper with a sparse grid printed in light blue or gray and right to the edge of the paper. This is often four squares to the inch for work not needing too much detail. It is sometimes referred to as *quadrille paper*.

Engineering paper is traditionally printed on light green or tan translucent paper. The grid lines are printed on the back side of each page and show through faintly to the front side. Each page has an unprinted margin. When photocopied or scanned, the grid lines typically do not show up in the resulting copy, which often gives the work a neat, uncluttered appearance. In the US and Canada, some engineering professors require student homework to be completed on engineering paper.

- Genkô yôshi – A type of manuscript paper used in Japan, normally printed with 400 squares in two sets of 20 lines of 10, used for compositions written in vertical script.
- Hexagonal paper shows regular hexagons instead of squares. These can be used to map geometric tiled or tesselated designs among other uses

Isometric graph paper or 3D graph paper is a triangular graph paper which uses a series of three guidelines forming a 60° grid of small triangles. The triangles are arranged in groups of six to make hexagons. The name suggests the use for isometric views or pseudo-three dimensional views. Among other functions, they can be used in the design of trianglepoint embroidery. It can be used to draw angles accurately,

- Logarithmic paper has rectangles drawn in varying widths corresponding to logarithmic scales for semilog graphs or log-log graphs.
- Normal probability paper is another graph paper with rectangles of variable widths. It is designed so that "the graph of the normal distribution function is represented on it by a straight line".

- Polar coordinate paper has concentric circles divided into small arcs or 'pie wedges' to allow plotting in polar coordinates.

In general, graphs showing grids are sometimes called Cartesian graphs because the square can be used to map measurements onto a Cartesian (x vs. y) coordinate system. It is also available without lines but with dots at the positions where the lines would intersect.

***Figure:** Isometric paper*

Graphic Communication

Graphic communication as the name suggests is communication using graphic elements. These elements include symbols such as glyphs and icons, images such as drawings and photographs, and can include the passive contributions of substrate, color and surroundings. It is the process of creating, producing, and distributing material incorporating words and images to convey data, concepts, and emotions.

The field of graphic communications encompasses all phases of the graphic communications processes from origination of the idea (design, layout, and typography) through reproduction, finishing and distribution of two- or three-dimensional products or electronic transmissions.

Overview

Graphic communication involves the use of visual material to relate ideas, such as drawings, photographs, slides, transparencies and sketches. The drawings of plans and refinements and a rough map sketched to show the way could be considered graphical communication.

Any medium that uses a graphics to aid in conveying a message, instruction, or an idea is involved in graphical communication. One of the most widely used forms of graphical communication is the drawing.

History

The earliest graphics known to anthropologists studying prehistoric periods are cave paintings and markings on boulders, bone, ivory, and antlers, which were created during the Upper Paleolithic period from 40,000–10,000 B.C. or earlier. Many of these were found to record astronomical, seasonal, and chronological details.

Some of the earliest graphics and drawings known to the modern world, from almost 6,000 years ago, are that of engraved stone tablets and ceramic cylinder seals, marking the beginning of the historic periods and the keeping of records for accounting and inventory purposes. Records from Egypt predate these and papyrus was used by the Egyptians as a material on which to plan the building of pyramids; they also used slabs of limestone and wood. From 600–250 BC, the Greeks played a major role in geometry. They used graphics to represent their mathematical theories such as the Circle Theorem and the Pythagorean theorem.

Graphic Communication Topics

Graphics

Graphics are visual presentations on some surface, such as a wall, canvas, computer screen, paper, or stone to brand, inform, illustrate, or entertain. Examples are photographs, drawings, Line Art, graphs, diagrams, typography, numbers, symbols, geometric designs, maps, engineering drawings, or other images. Graphics often combine text, illustration, and color. Graphic design may consist of the deliberate selection, creation, or arrangement of typography alone, as in a brochure, flier, poster, web site, or book without any other element. Clarity or effective communication may be the objective, association with other cultural elements may be sought, or merely, the creation of a distinctive style.

Graphics can be functional or artistic. The latter can be a recorded version, such as a photograph, or an interpretation by a scientist to highlight essential features, or an artist, in which case the distinction with imaginary graphics may become blurred.

Communication

Communication is the process whereby information is imparted by a sender to a receiver via a medium. It requires that all parties have an area of communicative commonality. There are auditory means, such as speaking, singing and sometimes tone of voice, and nonverbal, physical means, such as body language, sign language, paralanguage, touch, eye contact, by using writing. Communication is defined as a process by which we assign and convey meaning in an attempt to create shared understanding. This process requires a vast repertoire of skills in intrapersonal and interpersonal processing, listening, observing, speaking, questioning, analyzing, and evaluating. if you use these processes it is developmental and transfers to all areas of life: home, school, community, work, and beyond. It is through communication that collaboration and cooperation occur.

Visual Communication

Visual communication as the name suggests is communication through visual aid. It is the conveyance of ideas and information in forms that can be read or looked upon. Primarily associated with two dimensional images, it includes: signs, typography, drawing, graphic design, illustration, colour and electronic resources. It solely relies on vision. It is form of communication with visual effect. It explores the idea that a visual message with text has a greater power to inform, educate or persuade a person.

It is communication by presenting information through Visual form. The evaluation of a good visual design is based on measuring comprehension by the audience, not on aesthetic or artistic preference. There are no universally agreed-upon principles of beauty and ugliness. There exists a variety of ways to present information visually, like gestures, body languages, video and TV. Here, focus is on the presentation of text, pictures, diagrams, photos, et cetera, integrated on a computer display. The term visual presentation is used to refer to the actual presentation of information. Recent research in the field has focused on web design and graphically oriented usability. Graphic designers use methods of visual communication in their professional practice.

Communication Design

Communication design is a mixed discipline between and information-development which is concerned with how intermission such as printed, crafted, electronic media or presentations communicate with people. A communication design approach is not only concerned with developing the message aside from the aesthetics in media, but also with creating new media channels to ensure the message reaches the target audience. Communication design seeks to attract, inspire, create desires and motivate the people to respond to messages, with a view to making a favourable impact to the bottom line of the commissioning body, which can be either to build a brand, move sales, or for humanitarian purposes. Its process involves strategic business thinking, utilizing market research, creativity, and problem-solving.

Graphic Design

The term graphic design can refer to a number of artistic and professional disciplines which focus on visual communication and presentation. Various methods are used to create and combine symbols, images and/or words to create a visual representation of ideas and messages. A graphic designer may use typography, visual arts and page layout techniques to produce the final result. Graphic design often refers to both the process (designing) by which the communication is created and the products (designs) which are generated.

Common uses of graphic design include magazines, advertisements, product packaging and web design. For example, a product package might include a logo or other artwork, organized text and pure design elements such as shapes and color which unify the piece. Composition is one of the most important features of graphic design especially when using pre-existing materials or diverse elements.

Graphical Representation

The term representation, according to O'Shaughnessy and Stadler (2005), can carry a range of meanings and interpretations. In literary theory representation is commonly defined in three ways.

1. To look like or resemble
2. To stand in for something or someone
3. To present a second time to re-present

Representation, according to Mitchell (1995), began with early literary theory in the ideas of Plato and Aristotle, and has evolved into a significant component of language, Saussurian and communication studies. Aristotle discusses representation in three ways:

1. *The object:* The symbol being represented.
2. *Manner:* The way the symbol is represented.
3. *Means:* The material that is used to represent it.

The *means* of literary representation is language. The means of graphical representation are graphics. Graphical representation of data is one of the most commonly used modes of presentation.

House Plan

A house plan is a set of construction or working drawings (sometimes still called blueprints) that define all the construction specifications of a residential house such as dimensions, materials, layouts, installation methods and techniques.

Drawing Set

The principal information provided in set of house drawings are as follows:

- Site plans are drawn to show the location of a home on the property in its context. It is an overhead view of the construction site and the home as it sits in reference to the boundaries of the lot. Site plans should outline location of utility services, setback requirements, easements, location of driveways and walkways, and sometimes even topographical data that specifies the slope of the terrain.
- A floor plan is an overhead view of the completed house. You'll see parallel lines that scale at whatever width the walls are required to be. Dimensions are usually drawn between the walls to specify room sizes and wall lengths. Floor plans will also indicate rooms, all the doors and windows and any built-in elements, such as plumbing fixtures and cabinets, water heaters and furnaces, etc. Floor plans will include notes to specify finishes, construction methods, or symbols for electrical items.

- Elevations are a non-perspective view of the home. These are drawn to scale so that measurements can be taken for any aspect necessary. Plans include front, rear and both side elevations. The elevations specify ridge heights, the positioning of the final fall of the land, exterior finishes, roof pitches and other details that are necessary to give the home its exterior architectural styling.
- A section cuts through the dwelling and the location of this 'cut through' is noted on the floor plan. It describes how the building will be constructed and discusses how the internal finishes are to look. Sections are used because they explain certain conditions in more detail. These conditions may include ceiling height, ceiling type (flat or vault), and window and door dimensions.

Additional information as follows may also be provided:

- Foundation plan, including dimensions and locations for footings.
- Framing plan, for wall, including the size of the lumber to be used - usually 2x4 or 2x6.
- The Sub-floor Plan gives details of how this area will be constructed and how services will be arranged.
- Roof plans, including type, pitch and framing.
- Interior elevation drawings (interior walls).
- Detail drawings, such as built-in shelving, moldings, and columns.
- Schedules for elements such as windows and doors.
- Structural layouts.
- Electrical and telecoms drawings: show locations of all the outlets, TV sockets, switches and fixtures. Also indicates which switches operate which lights, and where the electric lines should be run,.
- Plumbing schematic drawing: plumbing fixtures and piping.

Lines and Symbols

House plans use the following lines and symbols to convey the relationship between objects:

- Dimension lines, which consist of a solid line with a mark at either end; space between the two marks equals the distance noted next to the line.

- Wall : one uses thick solid lines for walls.
- Partial height wall, low wall that doesn't go all the way to the ceiling.
- Thin solid lines are used for built-in structures (such as cabinets, bookshelves, or plumbing fixtures).
- Thin dotted lines indicate overhead features, such as wall cabinets in a kitchen or a special ceiling treatment or an archway in the living room.
- North arrow
- Doors:
  - o Door swing
  - o Bifold doors
  - o Passage (hinged) doors
- Sliding glass door
- Window
- Toilet
- Bathtub
- Shower
- Bathroom sink
- Kitchen sink
- Kitchen range
- Refrigerator
- Washer/Dryer
- Tree
- Shrub

Spaces and Rooms

House plans generally include the following features, depending on the size and type of house:

- Patios, terrace or veranda
- Garage
- Formal dining room
- Porch
- Bedroom
- Bathroom

- Kitchen
- Breakfast nook
- Living room
- Fireplace
- Basements
- Offices
- Closets
- Den (study)
- Attics
- Hallways, Vestibule
- and other interiors

Open Floorplan

An open floorplan is one which values wide, unenclosed spaces. They are made possible by curtain wall technology, engineered lumber and truss systems, and thoughtful structural design which allow for larger clear-spans than formerly possible in residential construction.

An open floor plan limits or excludes walls separating the main living areas of the house (living room, dining room, kitchen) to allow for entertaining or interaction between family.

Lesbian Rule

A lesbian rule was historically a flexible mason's rule made of lead that could be bent to the curves of a molding, and used to measure or reproduce irregular curves.

The rule is alluded to by Aristotle in his *Nicomachean Ethics* (bk V, ch. 10) as a metaphor for the importance of flexibility in equitable justice: "For what is itself indefinite can only be measured by an indefinite standard, like the leaden rule used by Lesbian builders; just as that rule is not rigid but can be bent to the shape of the stone, so a special ordinance is made to fit the circumstances of the case."

In the 17th and early 18th centuries the term was often used figuratively (as Aristotle had used it) to mean a pliant, flexible and accommodating principle of judgement. Thus Samuel Daniel in 1603 described equity as "that Lesbian square, that building fit, Plies to the worke, not forc'th the worke to it". In the later 17th century, the antiquary John Aubrey used the metaphor in a more pejorative sense, implying the distortion of evidence to fit a preconceived theory. He accused Inigo Jones, who had interpreted Stonehenge as a Roman monument, of having "made a Lesbians rule, which is conformed to the stone: that is, he framed the Monument to his own Hypothesis, which is much differing from the Thing it self".

Linear Scale

A linear scale, also called a bar scale, scale bar, graphic scale, or graphical scale, is a means of visually showing the scale of a map, nautical chart, engineering drawing, or architectural drawing.

On large scale maps and charts, those covering a small area, and engineering and architectural drawings, the linear scale can be very simple, a line marked at intervals to show the distance on the earth or object which the distance on the scale represents. A person using the map can use a pair of dividers (or, less precisely, two fingers) to measure a distance by comparing it to the linear scale. The length of the line on the linear scale is equal to the distance represented on the earth multiplied by the map or chart's scale.

In most projections, scale varies with latitude, so on small scale maps, covering large areas and a wide range of latitudes, the linear scale must show the scale for the range of latitudes covered by the map. One of these is shown below.

Since most nautical charts are constructed using the Mercator projection whose scale varies substantially with latitude, linear scales are not used on charts with scales smaller than approximately 1/ 80,000. Mariners generally use the nautical mile, which, because a nautical mile is approximately equal to a minute of latitude, can be measured against the latitude scale at the sides of the chart.

While linear scales are used on architectural and engineering drawings, particularly those that are drawn after the subject has been built, many such drawings do not have a linear scale and are marked "Do Not Scale Drawing" in recognition of the fact that paper size changes with environmental changes and only dimensions that are specifically shown on the drawing can be used reliably in precise manufacturing.

Nomenclature

The terms "bar scale", "graphic scale", "graphical scale", "linear scale", and "scale" are all used. *Bowditch* defined only "bar scale" in its 1962 Glossary, but added a reference to "graphic scale" by its 2002 edition. Dutton used both terms in 1978. The International

Hydrographic Organization's *Chart No. 1* uses only "linear scale". The British Admiralty's *Mariner's Handbook* uses "scale" to describe a linear scale and avoids confusion by using "natural scale" for the fraction defined at scale (map).

Lofting

Lofting is a drafting technique (sometimes using mathematical tables) whereby curved lines are generated, to be used in plans for streamlined objects such as aircraft and boats. The lines may be drawn on wood and the wood then cut for advanced woodworking. The technique can be as simple as bending a flexible object, such as a long strip of thin wood or thin plastic, so that it passes over three non-linear points and scribing the resultant curved line, or plotting the line using computers or mathematical tables.

Lofting is particularly useful in boat building, when it is used to draw and cut pieces for hulls and keels, which are usually curved, often in three dimensions. It was also commonly used in aircraft design before the widespread adoption of computer-generated shaping programmes.

Basic Lofting

As ship design evolved from craft to science, designers learned various ways to produce long curves on a flat surface. Generating and drawing such curves became a part of ship lofting; "lofting" means drawing full-sized patterns, so called because it was often done in large, lightly-constructed mezzanines or lofts above the factory floor. When aircraft design progressed beyond the stick-and-fabric boxes of its first decade of existence, the practice of lofting moved naturally into the aeronautical realm. As the storm clouds of World War II gathered in Europe, a US aircraft company, North American Aviation, took the practice into the purely mathematical realm. One of that war's outstanding warplanes, the North American P-51 Mustang, was designed using mathematical charts and tables rather than lofting tables.

Lofting is the transfer of a *Lines Plan* to a *Full Sized Plan.* This helps to assure that the boat will be accurate in its layout and pleasing in appearance. There are many methods to loft a set of plans.

Generally, boat building books have a detailed description of the lofting process, beyond the scope of this article. Plans can be lofted on a level wooden floor, marking heavy paper such as Red Rosin for the full-sized plans or directly on plywood sheets. The first step is to lay out the grid, mark the Base Line along the length of the paper or plywood sheet.

Then nail Battens every 12 inches (or more in some cases) where the station lines are to be set as a mark for the perpendicular line, which is marked with a T-square. The previous steps are followed in turn by marking the Top Line and the Water Line. Before continuing make sure to check the lines by using the Pythagorean theorem and make sure the grid is square.

The second step is to mark the points from the table of offsets. All measurements off the table of offsets are listed in Millimetres or the Feet, Inches and Eighths. The points are plotted at each station then use a small nail and a batten to Fair (draw with a fair curve) the boat's lines.

Definitions

- Full Sized Plan: a 1:1 scale construction drawing of a boat and its parts
- Lines Plan: a scaled-down version of a full-sized drawing often including the body, plan, profile and section views
- Body Plan is a view of the boat from both dead ahead and dead astern split in half
- Plan view is looking down on the boat from the top
- Profile View is a view of the boat from the side
- Section View Cross section of the boats width
- Batten long stick to help draw fair lines

Pantograph

A pantograph is a mechanical linkage connected in a manner based on parallelograms so that the movement of one pen, in tracing an image, produces identical movements in a second pen.

If a line drawing is traced by the first point, an identical, enlarged, or miniaturized copy will be drawn by a pen fixed to the other. Using the same principle, different kinds of pantographs are used for other forms of duplication in areas such as sculpture, minting, engraving and milling.

Because of the shape of the original device, a pantograph also refers to a kind of structure that can compress or extend like an accordion, forming a characteristic rhomboidal pattern. This can be found in extension arms for wall-mounted mirrors, temporary fences, scissor lifts, and other scissor mechanisms such as the pantograph used in electric locomotives and trams.

History

The first pantograph was constructed in 1603 by Christoph Scheiner, who used the device to copy and scale diagrams, but he wrote about the invention over 27 years later, in *"Pantographice"* (Rome 1631). One arm of the pantograph contained a small pointer, while the other held a drawing implement, and by moving the pointer over a diagram, a copy of the diagram was drawn on another piece of paper. By changing the positions of the arms in the linkage between the pointer arm and drawing arm, the scale of the image produced can be changed. A more complicated version called the eidograph was developed by William Wallace (1768–1843) in 1831.

Uses

Drafting

The original use of the pantograph was for copying and scaling line drawings. Modern versions are sold as toys.

History

The first pantograph was constructed in 1603 by Christoph Scheiner, who used the device to copy and scale diagrams, but he wrote about the invention over 27 years later, in *"Pantographice"* (Rome 1631). One arm of the pantograph contained a small pointer, while the other held a drawing implement, and by moving the pointer over a diagram, a copy of the diagram was drawn on another piece of paper. By changing the positions of the arms in the linkage between the pointer arm and drawing arm, the scale of the image produced can be changed. A more complicated version called the eidograph was developed by William Wallace (1768–1843) in 1831.

Uses

Drafting

The original use of the pantograph was for copying and scaling line drawings. Modern versions are sold as toys.

Before the advent of control technologies such as numerical control (NC and CNC) and programmable logic control (PLC), duplicate parts being milled on a milling machine could not have their contours mapped out by moving the milling cutter in a "connect-the-dots" ("by-the-numbers") fashion. The only ways to control the movement of the cutting tool were to dial the positions by hand using dexterous skill (with natural limits on a human's accuracy and precision) or to trace a cam, template, or model in some way, and have the cutter mimic the movement of the tracing stylus. If the milling head was mounted on a pantograph, a duplicate part could be cut (and at various scales of magnification besides 1:1) simply by tracing a template. (The template itself was usually made by a tool and die maker using toolroom methods, including milling via dialing followed by hand sculpting with files and/or die grinder points.) This was essentially the same concept as reproducing documents with a pen-equipped pantograph, but applied to the machining of hard materials such as metal, wood, or plastic. Pantograph routing, which is conceptually identical to pantograph milling, also exists (as does CNC routing).

The development and dissemination throughout industry of NC, CNC, PLC, and other control technologies provided a new way to control the movement of the milling cutter: via feeding information from a programme to actuators (servos, selsyns, leadscrews, machine slides, spindles, and so on) that would move the cutter as the information directed. Today most commercial machining is done via such programmable, computerized methods. Home machinists are likely to work via manual control, but computerized control has reached the home-shop level as well (it's just not yet as pervasive as its commercial counterparts). Thus pantograph milling machines are largely a thing of the past. They are still in commercial use, but at a greatly reduced and ever-dwindling level. They are no longer built new by machine tool builders, but a small market for used machines still exists. As for the magnification-and-reduction feature of a pantograph (with the scale determined by the adjustable arm lengths), it is achieved in CNC via mathematic calculations that the computer applies to the programme information practically instantaneously. Scaling functions (as well as mirroring functions) are built into languages such as G-code.

Other Uses

Perhaps the pantograph that is most familiar to the general public is the extension arm of an adjustable wall-mounted mirror.

In another application similar to drafting, the pantograph is incorporated into a pantograph engraving machine with a revolving cutter instead of a pen, and a tray at the pointer end to fix precut lettered plates (referred to as 'copy'), which the pointer follows and thus the cutter, via the pantograph, reproduces the 'copy' at a ratio to which the pantograph arms have been set.

The typical range of ratio is Maximum 1:1 Minimum 50:1 (reduction) In this way machinists can neatly and accurately engrave numbers and letters onto a part.

The device which maintains electrical contact with the contact wire and transfers power from the wire to the traction unit, used in electric locomotives and trams, is also called a "pantograph".

Some types of trains on the New York City Subway use end pantograph gates (which, to avoid interference, compress under spring pressure around curves while the train is en route) to prevent passengers on station platforms from falling into or riding in the gaps between the cars.

Old-style 'baby gates' used a 2-dimensional pantograph mechanism (in a similar style to pantograph gates on subway cars) as a means of keeping toddlers away from stairways. The openings in these gates are too large to meet modern baby gate safety standards.

Herman Hollerith's "Keyboard punch" used for the 1890 U.S. Census was a pantograph design and sometimes referred to as "The Pantograph Punch".

An early 19th-century device employing this mechanism is the polygraph, which produces a duplicate of a letter as the original is written.

Longarm quilting machine operators may trace a pantograph, paper pattern, with a laser pointer to stitch a custom pattern onto the quilt. Digitized pantographs are followed by computerized machines.

Linn Boyd Benton invented a pantographic engraving machine for type design, which was capable not only of scaling a single font design pattern to a variety of sizes, but could also condense, extend, and slant the design (mathematically, these are cases of affine transformation, which is the fundamental geometric operation of most systems of digital typography today, including PostScript).

Pantographs are also used as guide frames in heavy-duty applications including scissor lifts, material handling equipment, stage lifts and speciality hinges (such as for panel doors on boats and airplanes).

Richard Feynman used the analogy of a pantograph as a way of scaling down tools to the nanometre scale in his talk There's Plenty of Room at the Bottom.

Numerous trade-show displays use 3-dimensional pantograph mechanisms to support backdrops for exhibit booths. The framework expands in 2 directions (vertical and horizontal) from a bundle of connected rods into a self-supporting structure on which a fabric backdrop is hung.

Perspectivity

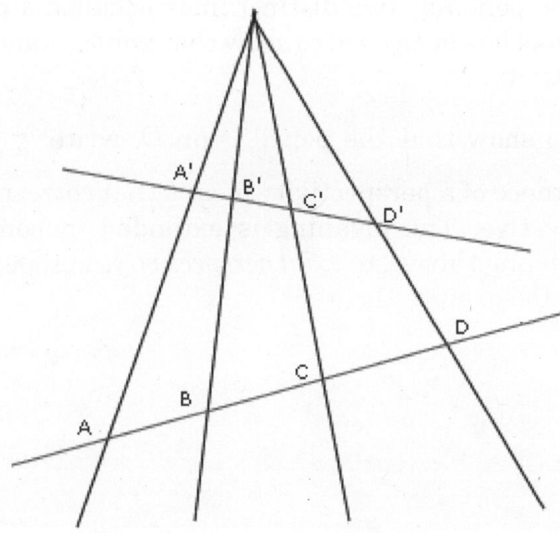

A perspectivity:

$$ABCD \doublebarwedge A'B'C'D',$$

In geometry and in its applications to drawing, a perspectivity is the formation of an image in a picture plane of a scene viewed from a fixed point. The science of graphical perspective uses perspectivities to make realistic images in proper proportion. According to Kirsti Andersen, the first author to describe perspectivity was Leon Alberti in his *De Pictura* (1435). In English, Brook Taylor presented his *Linear Perspective* in 1715, where he explained "Perspective is the Art of drawing on a Plane the Appearances of any figures, by the Rules of Geometry". In a second book, *New Principles of Linear Perspective* (1719), Taylor wrote

When Lines drawn according to a certain Law from the several Parts of any figure, cut a Plane, and by that Cutting or Intersection describe a figure on that Plane, that figure so described is called the *Projection* of the other figure. The Lines producing that Projection, taken all together, are called the *System of Rays.* And when those Rays all pass thro' one and same Point, they are called the *Cone of Rays.* And when that Point is consider'd as the Eye of a Spectator, that System of Rays is called the *Optic Cone*

In projective geometry the points of a line are called a projective range, and the set of lines on a point is called a pencil. According to H. S. M. Coxeter, "The correspondence between two ranges that are sections of one pencil by two distinct lines is called a perspectivity." A special symbol has been used to show that points X and Y are related by a perspectivity:

$X \overset{=}{\wedge} Y$. To show that the pencil is on O, write $X \overset{O}{\overset{=}{\wedge}} Y$.

The existence of a perspectivity means that corresponding points are in perspective. The meaning is extended, beyond the *central perspectivity* defined above, to *axial perspectivity*, a property of triangles expressed in Desargues' theorem.

Protractor

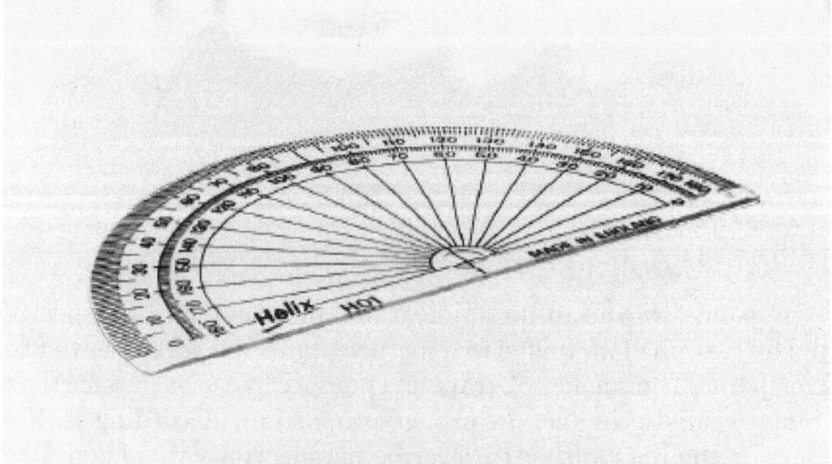

Figure: *A half circle protractor marked in degrees (180°).*

A protractor is a square, circular or semicircular tool, typically made of transparent plastic or glass, for measuring angles. Most protractors measure angles in degrees (°). Radian-scale protractors measure angles in radians.

They are used for a variety of mechanical and engineering-related applications, but perhaps the most common use is in geometry lessons in schools.

Some protractors are simple half-discs. More advanced protractors, such as the bevel protractor, have one or two swinging arms, which can be used to help measure the angle.

Bevel Protractor

A bevel is——; used for measuring or marking off angles. Sometimes Vernier scales are attached to give more precise readings. It has wide application in architectural and mechanical drawing, although its use is decreasing with the availability of modern drawing software or CAD.

Universal bevel protractors are also used by toolmakers; as they measure angles by mechanical contact they are classed as mechanical protractors.

The bevel protractor is used to establish and test angles to very close tolerances. It reads to 5 minutes or 1/12° and can measure any angle from 0° to 360°.

The bevel protractor consists of a beam, a graduated dial and a blade which is connected to a swivel plate (with Vernier scale) by thumb nut and clamp. When the edges of the beam and blade are parallel, a small mark on the swivel plate coincides with the zero line on the graduated dial. To measure an angle between the beam and the blade of 90° or less, the reading may be obtained direct from the graduation number on the dial indicated by the mark on the swivel plate. To measure an angle of over 90°, subtract the number of degrees as indicated on the dial from 180°, as the dial is graduated from opposite zero marks to 90° each way.

Since the spaces, both on the main scale and the Vernier scale, are numbered both to the right and to the left from zero, any angle can be measured. The readings can be taken either to the right or to the left, according to the direction in which the zero on the main scale is moved.

The above picture illustrates a variety of uses of the bevel protractor.

Reading the Vernier scale: The bevel protractor Vernier scale may have graduations of 52 (minutes) or 1/12°. Each space on the Vernier scale is 52 less than two spaces on the main scale. Twenty four spaces on the Vernier scale equal in extreme length twenty three double degrees. Thus the difference between the space occupied by 2° on a main scale and the space of the Vernier scale is equal to one twenty-fourth of 2°, or 52 .

Read off directly from the main scale the number of whole degrees between 0 on this scale and the 0 of the Vernier scale. Then count, in the same direction, the number of spaces from the zero on the Vernier scale to a line that coincides with a line on the main scale; multiply this number by 5 and the product will be the number of minutes to be added to the whole number of degrees.

For example: Zero on the vernier scale has moved 28 whole degrees to the right of the 0 on the main scale and the 3rd line on the vernier scale coincides with a line upon the main scale as indicated. Multiplying 3 by 5, the product, 15, is the number of minutes to be added to the whole number of degrees, thus indicating a setting of 28 degrees and 15 minutes.

7

Artistic Rendering

Rendering in visual art and technical drawing means the process of formulating, adding color, shading, and texturing of an image. It can also be used to describe the quality of execution of that process. When used as a means of expression, it is synonymous with illustrating. However, it may be used for mere visualization of existing data regardless of any preconceived message or idea to express.

Rendering style

Artistic rendering is the application of visual art styles to rendering. For photorealistic rendering styles, the emphasis is on accurate reproduction of light-and-shadow and the surface properties of the depicted objects, composition, or other more generic qualities. When the emphasis is on unique interpretive rendering styles, visual information is interpreted by the artist and displayed accordingly using the chosen art medium and level of abstraction in abstract art. In computer graphics, interpretive rendering styles are known as non-photorealistic rendering styles, but may be used to simplify technical illustrations. Rendering styles that combine photorealism with non-photorealism are known as hyperrealistic rendering styles.

Ruling Pen

A ruling pen is a drawing instrument for drawing with ink or with other drawing fluids. A ruling pen contains ink in a slot between two flexible metal jaws, which are tapered to a point. It enables precise rendering of the thinnest lines. The line width can be adjusted by an adjustment screw connecting the jaws. The adjustment screw can optionally have a number dial.

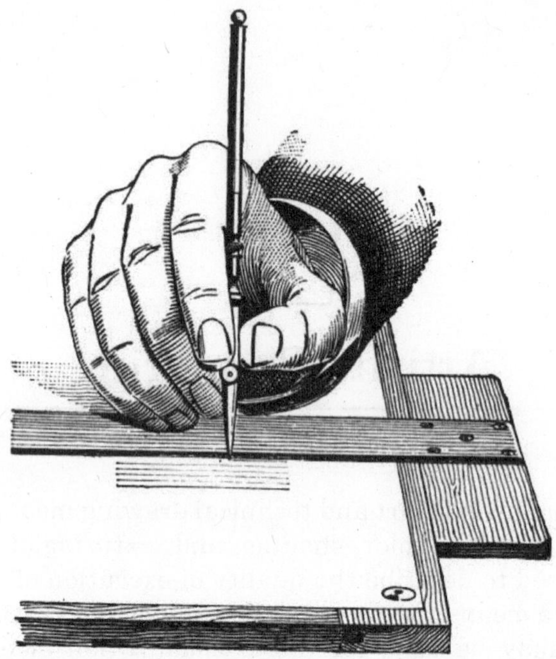

Figure: *Illustration of ruling pen use from* A Textbook on Ornamental Design *(1901)*

Originally used for technical drawings in engineering and cartography together with straight rulers and French curves, it is today used for specific uses, such as picture framing or calligraphy.

Set Square

In American English, a set square is an alternative name for a T-square.

Figure: *triangle, set square*

A set square or triangle (American English) is an object used in engineering and technical drawing, with the aim of providing a straightedge at a right angle or other particular planar angle to a baseline.

The most simple form of set square is a triangular piece of transparent plastic (or formerly of polished wood) with the centre removed. More commonly the set square bears the markings of a ruler and a half circle protractor. The outer edges are typically bevelled. These set squares come in two usual forms, both right triangles: one with 90-45-45 degree angles, the other with 30-60-90 degree angles. Combining the two forms by placing the hypotenuses together will also yield 15° and 75° angles. They are often purchased in packs with protractors and compasses.

Less commonly found is the adjustable set square. Here, the body of the object is cut in half and rejoined with a hinge marked with angles. Adjustment to the marked angle will produce any desired angle up to a maximum of 180°.

Shop Drawing

A shop drawing is a drawing or set of drawings produced by the contractor, supplier, manufacturer, subcontractor, or fabricator. Shop drawings are typically required for prefabricated components. Examples of these include: elevators, structural steel, trusses, pre-cast, windows, appliances, cabinets, air handling units, and millwork. Also critical are the installation and coordination shop drawings of the MEP trades such as sheet metal ductwork, piping, plumbing, fire protection, and electrical.

Shop drawings are not produced by architects and engineers under their contract with the owner. The shop drawing is the manufacturer's or the contractor's drawn version of information shown in the construction documents.

The shop drawing normally shows more detail than the construction documents. It is drawn to explain the fabrication and/or installation of the items to the manufacturer's production crew or contractor's installation crews. The style of the shop drawing is usually very different from that of the architect's drawing. The shop drawing's primary emphasis is on the particular product or installation and excludes notation concerning other products and installations, unless integration with the subject product is necessary.

Information Required to be Included in Shop Drawings

Comparison Information for the Architect and Engineer

The shop drawings should include information for the architect and engineer to compare to the specifications and drawings. The shop drawing should address the appearance, performance, and prescriptive descriptions in the specifications and construction drawings.

The shop drawing often is more detailed than the information shown in the construction documents to give the architect and engineer the opportunity to review the fabricator's version of the product, prior to fabrication. References to the construction documents, drawings, and specifications assist the architect and engineer in their review of the shop drawings. Attachment of manufacturer's material specifications, "catalog cut sheets," and other manufacturer's information may be helpful to accompany these drawings. Because shop drawings facilitate the architect's and engineer's approval of the product, they should be as clear and complete as possible.

Notes of Changes or Alterations from the Construction Documents

Notes concerning changes or differences from the original documents should be made on the shop drawing for the architect's and engineer's approval. Ultimately, they are responsible for changes in these drawings and should have the opportunity to analyze any modifications. A dialogue should occur between the fabricator and the architect and engineer about any areas needing clarification. Successful installations are the result of collaboration between the designer, fabricator, and contractor.

Information Needed to Fabricate the Product

Dimensions, manufacturing conventions, and special fabrication instructions should be included on the shop drawing. It should be clear to fabrication personnel what will be manufactured from the shop drawings alone. The construction documents are rarely used as a reference in fabrication, with the fabricators relying on the shop drawing for all information.

Indication of Dimensions Needing Verification from the Jobsite

Most jobsite dimensions, such as the dimensions between two surfaces on the jobsite, need to be verified. A dimension may be shown on the construction drawings, but the actual dimension may vary, from very small to large increments, depending on jobsite conditions.

It is extremely important that the fabricated item arrive on the jobsite ready to be installed without field modification. Special care must be taken by the contractor to measure and verify dimensions. In new construction, plan dimensions usually are sufficient for ordering many fabricated items such as structural steel or precast concrete.

In remodelling and renovation work, it is essential that field dimensions be verified prior to fabrication. Some fabricators, such as cabinet and casework suppliers, prefer not to rely on the contractor's verification and will verify the dimensions with their own personnel.

Placement or Installation information

Some fabricators and manufacturers will provide symbols, data, or instructions concerning installation. This can include a list of other materials, such as fasteners or adhesives, appropriate but not included for the product.

Third party review may be required for major building systems. An example of this would be a commercial chiller which would be furnished by the mechanical contractor, but would require electrical connections, plumbing, rigging, insulation and commissioning. Various third-parties will need to review the installation information and confirm they are furnishing compatible equipment and proper layout of services. Review of installation information for major equipment should be reviewed with field supervisors including the project superintendent, trade foremen and field engineer. Installation of major equipment will dictate structural clearances and temporary openings.

Samples

Some fabrications will require a sample submittal with the shop drawing, primarily for color and texture selection of finishes.

Computer-aided Mechanical, Electrical and Plumbing Design Coordination

Problems with design coordination, such as time consumption and ineffectiveness related to the current 2D paper-based process, are some of the top concerns of a general contractor, since late conflict correction increases the potential for errors in the field. In order to address this issue, there have been an increasing number of Architecture, Engineering and Construction (AEC) firms utilizing building information models (BIM) in their coordination and clash detection processes, which according to practitioners, allows for increased coordination and fewer field problems. Leite et al. compared types of clashes identified in a manual coordination process (overlay of 2D drawings on a light table by pairs of subcontractors) and through automatic clash detection using a Building Information Model (BIM). The automatic clash detection identified several clashes that were missed by the subcontractors, who were performing this task manually.

Also, the manual clash detection identified clashes which could not possibly be found by the automatic clash detection software, since one of the clashing objects (e.g. cable trays) was not modelled in the BIM. This study also included site observations of field detected clashes, some of which were not identified in either manual or automatic processes.

Leite noted that the combination of clashes identified in coordination meetings, those automatically detected, as well as those identified in the field enable identification of objects that need to be modelled in order to capture the largest possible number of clashes. This paper was limited to the comparison of types of clashes identified in each of the three methods during a specific project. Although their results cannot be generalized, they still provide insight towards the need to identify what needs to be modelled in a BIM for MEP coordination prior to the start of the coordination process.

Reviews

Because writing comments on eight to ten copies is a tedious process and an inefficient use of the architect and engineer's time, they will frequently specify other methods for distributing comments. Quick review is essential during the approval process. Any method that facilitates this, while providing ample opportunity for comment and complete distribution, should be considered. Although a procedure may be specified in the contract drawings, most architects and engineers are open to suggestions and innovations that speed up the process.

Number of Copies

Shop drawings are required, in various forms, depending upon the practice of the architect and engineer. A specific number of copies may be required by the specification. An example distribution of the completed and corrected shop drawings may include the:

- Owner—file or inspection copy
- Architect—file copy
- Architect—field copy or inspection copy
- Consulting engineer—file copy

- Consulting engineer—inspection copy
- Contractor—file copy
- Contractor—field copy
- Supplier—original copy or one copy

Submittal of one or two Copies of the shop Drawing

Corrections are made by the architect and engineer, and the shop drawing is corrected by the supplier, then the appropriate number of copies is distributed. This method can be time consuming, as the shop drawing is not approved until the corrections are made on it.

Submittal of a Copy that can be Reproduced

The architect and engineer make comments on the reproducible, then copies are distributed. This method facilitates the timely approval and distribution of the shop drawing. Review comments usually are obvious on the reproducible copy. When sepia copies are used, the reproduction of the sepia often is not as clear as a normal blue-line print.

Compatible CAD Software

When the supplier and designer have compatible CAD software or when universal file formats such as IFC, PDF or DWG are utilized, the review can be made from a CD, email or FTP transfer. Comments can be made by the designer in a bold font or changes can be boxed for emphasis.

Popular CAD platforms used for generating shop drawings are Advance Steel, AutoCAD, CATIA, Creo Elements/Pro, Inventor, Solidworks and Tekla Structures.

Shop Drawings in Concrete Reinforcing

Concrete reinforcing is one of the many items requiring specialized shop drawings for the fabrication of the material. Concrete reinforcing is custom-fabricated from 60-foot-long reinforcing bars. The reinforcing bars are cut to length and bent to specific configurations. The shop drawing and the accompanying "cut sheet" lists the quantity, sizes, lengths, and shapes of the reinforcing bar. This information is provided for review by the structural engineer to ensure that sufficient reinforcing is being supplied; fabrication of the bar by the supplier's shop; an inventory list for the contractor, upon delivery the typical project has thousands of pieces of reinforcing steel that need to be organized for storage and installation; and placement by the ironworker.

The Concrete Reinforcing Steel Institute (CRSI) has developed standard symbols, graphics, and formats for shop drawings and cut sheets that generally are used by reinforcing steel fabricators. Each fabricator, has particular style for shop drawings and cut sheets, depending on the draftspeople and Computer-aided design systems. Examples of software used are Advance Concrete, AutoCAD, MicroStation, ProConcrete and Tekla Structures.

Specification (Technical Standard)

Specification (often abbreviated as spec) may refer to an explicit set of requirements *to be satisfied* by a material, design, product, or service. Should a material, product, or service fail to meet one or more of the applicable specifications, it may be referred to as being *out of specification*; the abbreviation OOS may also be used. In casual usage, *underspec* or *overspec* are used when something is worse or better than specified (compare overengineering), though in general (such as for sizes) there is only a notion of "in spec" or "out of spec", not "better" or "worse". A specification is a type of technical standard.

A technical specification may be developed by any of various kinds of organizations, both public and private. Example organization types include a corporation, a consortium (a small group of corporations), a trade association (an industry-wide group of corporations), a national government (including its military, regulatory agencies, and national laboratories and institutes), a professional association (society), a purpose-made standards organization such as ISO, or vendor-neutral developed generic requirements (Telcordia Technologies). It is common for one organization to *refer to* (*reference*, *call out*, *cite*) the standards of another. Voluntary standards may become mandatory if adopted by a government or business contract.

A Design or Product Specification describes the features of the solutions for the Requirement Specification. Sometimes the term *specification* is here used in connection with a data sheet (or *spec sheet*). This may be confusing. A data sheet describes the technical characteristics of an item or product as designed and/or produced. It can be published by a manufacturer to help people choose products or to help use the products. A data sheet is not a technical specification as described in this article.

Use

In engineering, manufacturing, and business, it is vital for suppliers, purchasers, and users of materials, products, or services to understand and agree upon all requirements. A specification is a type of a standard which is often referenced by a contract or procurement document. It provides the necessary details about the specific requirements.

Specifications may be written by government agencies, standards organizations (ASTM, ISO, CEN, DoD, etc.), trade associations, corporations, and others.

A product specification does not necessarily prove a product to be correct or useful. An item might be verified to comply with a specification or stamped with a specification number: This does not, by itself, indicate that the item is fit for any particular use. The people who use the item (engineers, trade unions, etc.) or specify the item (building codes, government, industry, etc.) have the responsibility to consider the choice of available specifications, specify the correct one, enforce compliance, and use the item correctly. Validation of suitability is necessary.

Guidance and Content

Sometimes a guide or a standing operating procedure is available to help write and format a good specification. A specification might include:

- Descriptive title, number, identifier, etc. of the specification
- Date of last effective revision and revision designation
- A logo or trademark to indicate the document copyright, ownership and origin
- Table of Contents (TOC), if the document is long
- Person, office, or agency responsible for questions on the specification, updates, and deviations.
- The significance, scope or importance of the specification and its intended use.
- Terminology, definitions and abbreviations to clarify the meanings of the specification
- Test methods for measuring all specified characteristics
- Material requirements: physical, mechanical, electrical, chemical, etc. Targets and tolerances.
- Acceptance testing, including Performance testing requirements. Targets and tolerances.
- Drawings, photographs, or technical illustrations

- Workmanship
- Certifications required.
- Safety considerations and requirements
- Environmental considerations and requirements
- Quality control requirements, acceptance sampling, inspections, acceptance criteria
- Person, office, or agency responsible for enforcement of the specification.
- Completion and delivery.
- Provisions for rejection, reinspection, rehearing, corrective measures
- References and citations for which any instructions in the content maybe required to fulfill the traceability and clarity of the document
- Signatures of approval, if necessary
- Change record to summarize the chronological development, revision and completion if the document is to be circulated internally
- Annexes and Appendices that are expand details, add clarification, or offer options.

Process Capability Considerations

A good engineering specification, by itself, does not necessarily imply that all products sold to that specification actually meet the listed targets and tolerances. Actual production of any material, product, or service involves inherent variation of output. With a normal distribution, the tails of production may extend well beyond plus and minus three standard deviations from the process average.

The process capability of materials and products needs to be compatible with the specified engineering tolerances. Process controls must be in place and an effective Quality management system, such as Total Quality Management, needs to keep actual production within the desired tolerances. Effective enforcement of a specification is necessary for it to be useful.

Construction Specifications in North America

Specifications in North America form part of the contract documents that accompany and govern the construction of a building.

The guiding master document is the latest edition of MasterFormat. It is a consensus document that is jointly sponsored by two professional organisations: Construction Specifications Canada and Construction Specifications Institute.

While there is a tendency to believe that "Specs overrule Drawings" in the event of discrepancies between the text document and the drawings, the actual intent—made explicit in the contract between the Owner and the Contractor—is for the drawings and specifications to be complementary, together providing the information required for a complete facility.

The Specifications fall into 50 Divisions, or broad categories of work results involved in construction. The Divisions are subdivided into Sections, each one addressing a narrow scope of the construction work. For instance, firestopping is addressed in Section 078400 - Firestopping. It forms part of Division 07, which is Thermal and Moisture Protection. Division 07 also addresses building envelope and fireproofing work.

Each Section is subdivided into three distinct Parts: "General", "Products" and "Execution". The MasterFormat system can be successfully applied to residential, commercial, civil, and industrial construction.

Specifications can be either "performance-based", whereby the specifier restricts the text to stating the performance that must be achieved by the completed work, or "prescriptive", whereby the specifier indicates specific products, vendors and even contractors that are acceptable for each workscope. Most construction specifications are a combination of performance-based and prescriptive types, naming acceptable manufacturers and products while also specifying certain standards and design criteria that must be met.

While North American specifications are usually restricted to broad descriptions of the work, European ones can include actual work quantities, including such things as area of drywall to be built in square metres, like a bill of materials. This type of specification is a collaborative effort between a specwriter and a quantity surveyor. This approach is unusual in North America, where each bidder performs a quantity survey on the basis of both drawings and specifications.

Although specifications are usually issued by the architect's office, specwriting itself is undertaken by the architect and the various engineers or by specialist specwriters. Specwriting is often a distinct professional trade, with professional certifications such as "Certified Construction Specifier" (CCS) through the professional organizations noted above. Specwriters are either employees of or sub-contractors to architects, engineers, or construction management companies. Specwriters frequently meet with manufacturers of building materials who seek to have their products specified on upcoming construction projects so that contractors can include their products in the estimates leading to their proposals.

Construction Specifications in the UK

Specifications in the UK are part of the contract documents that accompany and govern the construction of a building. They are prepared by construction professionals such as Architects, Architectural technologists, Structural Engineers, Landscape Architects and Building Services Engineers. They are created from previous project specifications, in-house documents or master specifications such as the National Building Specification (NBS).

The National Building Specification is owned by the Royal Institute of British Architects (RIBA) through their commercial group RIBA Enterprises (RIBAe). NBS master specifications provide content that is broad and comprehensive, and delivered using software functionality that enables specifiers to customize the content to suit the needs of the project and to keep up to date. UK project specification types fall into two main categories prescriptive and performance.

Prescriptive specifications define the requirements using generic or proprietary descriptions of what is required, whereas performance specifications focus on the outcomes rather than the characteristics of the components. Specifications are an integral part of Building Information Modelling and cover the non-geometric requirements.

Food and Drug Specifications

Pharmaceutical products can usually be tested and qualified by various Pharmacopoeia. Current existing pronounced standards include:

- British Pharmacopoeia
- European Pharmacopoeia
- Japanese Pharmacopoeia
- The International Pharmacopoeia
- United States Pharmacopoeia

If any pharmaceutical product is not covered by the above standards, it can be evaluated by the additional source of Pharmacopoeia from other nations, from industrial specifications, or from a standardized formulary such as

- British National Formulary for Children
- British National Formulary
- National Formulary

A similar approach is adopted by the food manufacturing, of which Codex Alimentarius ranks the highest standards, followed by regional and national standards.

The coverage of food and drug standards by ISO is currently less fruitful and not yet put forward as an urgent agenda due to the tight restrictions of regional or national constitution

Specifications and other standards can be externally imposed as discussed above, but also intenal manufacturing and quality specifications. These exist not only for the food or pharmaceutical product but also for the processing machinery, quality processes, packaging, logistics (cold chain), etc. and are exemplified by *ISO 14134* and *ISO 15609* The converse of explicit statement of specifications is a process for dealing with observations that are out-of-specification. The United States Food and Drug Administration has published a non-binding recommendation that addresses just this point.

At the present time, much of the information and regulations concerning food and food products remain in a form which makes it difficult to apply automated information processing, storage and transmission methods and techniques.

Data systems that can process, store and transfer information about food and food products need formal specifications for the representations of data about food and food products in order to operate effectively and efficiently.

Development of formal specifications for food and drug data with the necessary and sufficient clarity and precision for use specifically by digital computing systems have begun to emerge from government agencies and standards organizations.

The United States Food and Drug Administration has published specifications for a "Structured Product Label" which drug manufacturers must by mandate use to submit electronically the information on a drug label.

Recently, ISO has made some progress in the area of food and drug standards and formal specifications for data about regulated substances through the publication of *ISO 11238*

Information Technology

Specification Need

Specifications are needed to avoid errors due to lack of specification, for instance, in interoperability issues.

For instance, when two applications share Unicode data, but use different normal forms or use them incorrectly, in an incompatible way or without sharing a minimum set of interoperability specification, errors and data loss can result. For example, Mac OS X has many components that prefer or require only decomposed characters (thus decomposed-only Unicode encoded with UTF-8 is also known as "UTF8-MAC"). In one specific instance, the combination of OS X errors handling composed characters, and the samba file- and printer-sharing software (which replaces decomposed letters with composed ones when copying file names), has led to confusing and data-destroying interoperability problems.

Applications may avoid such errors by preserving input code points, and only normalizing them to the application's preferred normal form for internal use.

Such errors may also be avoided by with algorithms normalizing both strings before any binary comparison.

However errors due to filename encoding incompatibilities have always existed, due to a lack of minimum set of common specification between software hoped to be interoperable between various filesystem drivers, operatings systems, network protocols, and thouthands software packages.

Formal Specification

A formal specification is a mathematical description of software or hardware that may be used to develop an implementation. It describes *what* the system should do, not (necessarily) *how* the system should do it. Given such a specification, it is possible to use formal verification techniques to demonstrate that a candidate system design is correct with respect to that specification. This has the advantage that incorrect candidate system designs can be revised before a major investment has been made in actually implementing the design. An alternative approach is to use provably correct refinement steps to transform a specification into a design, and ultimately into an actual implementation, that is correct by construction.

Programme Specification

A programme specification is the definition of what a computer programme is expected to do. It can be *informal*, in which case it can be considered as a user manual from a developer point of view, or *formal*, in which case it has a definite meaning defined in mathematical or programmatic terms. In practice, many successful specifications are written to understand and fine-tune applications that were already well-developed, although safety-critical software systems are often carefully specified prior to application development. Specifications are most important for external interfaces that must remain stable.

Functional Specification

In software development, a functional specification (also, functional spec or specs or functional specifications document (FSD)) is the set of documentation that describes the behaviour of a computer programme or larger software system. The documentation typically describes various inputs that can be provided to the software system and how the system responds to those inputs.

Web Service Specification

Further information: List of Web service specifications

Web services specifications are often under the umbrella of a quality management system.

Document Specification

These types of documents define how a specific document should be written, which may include, but is not limited to, the systems of a document naming, version, layout, referencing, structuring, appearance, language, copyright, hierarchy or format, etc. Very often, this kind of specifications is complemented by a designated template.

Straightedge

A straightedge is a tool with an edge free from curves, or straight, used for transcribing straight lines, or checking the straightness of lines. If it has equally spaced markings along its length it is usually called a ruler. Straightedges are used in the automotive service and machining industry to check the flatness of machined mating surfaces.

True straightness can in some cases be checked by using a laser line level as an optical straightedge: it can illuminate an accurately straight line on a flat surface such as the edge of a plank or shelf.

Compass-and-straightedge Construction

An idealized straightedge is used in compass-and-straightedge constructions in plane geometry. It may be used:

- Given two points, to draw the line connecting them.
- Given a point and a circle, to draw either tangent.
- Given two circles, to draw any of their common tangents.

It may not be marked or used together with the compass so as to transfer the length of one segment to another.

It is possible to do all compass and straightedge constructions without the straightedge. That is, it is possible, using only a compass, to find the intersection of two lines given two points on each, and to find the tangent points to circles. It is not, however, possible to do all constructions using only a straightedge. It *is* possible to do them with straightedge alone given one circle and its centre.

T-square

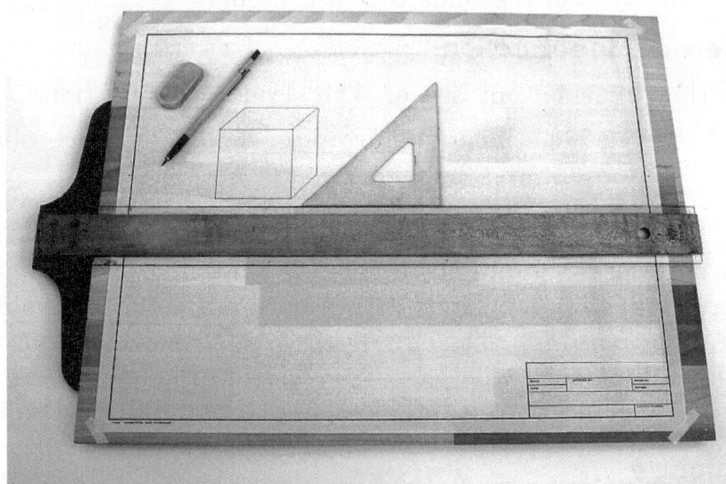

***Figure:** Drafting board with a T-square and triangle*

A T-square is a technical drawing instrument used by draftsmen primarily as a guide for drawing horizontal lines on a drafting table. It may also guide a set square to draw vertical or diagonal lines. Its name comes from its resemblance to the letter T.

T-squares come in varying sizes, common lengths being 18 inches (460 mm), 24 inches (610 mm), 30 inches (760 mm), 36 inches (910 mm) and 42 inches (1,100 mm).

Figure: *An old T-square.*

T-squares are also used to measure and cut drywall. Drywall T-squares are typically made of aluminium and have a 48-inch (1,200 mm) long tongue.

Higher-end table saws are often equipped with T-square fences. These table saw fences are attached to a rail only on the front side of the table unlike traditional fences which clamp to both the front and the rear of the table.

A T-square has two components - the long shaft called the "blade" and the short shaft called the "stock". The T-square usually has a transparent edge made of plastic which should be free of nicks and cracks in order to provide smooth, straight lines.

Ruler

A ruler, sometimes called a rule or line gauge, is an instrument used in geometry, technical drawing, printing as well as engineering and building to measure distances or to rule straight lines. The ruler is a straightedge which may also contain calibrated lines to measure distances.Types Rulers have long been made of many materials in a wide range of sizes. Some are wooden. Plastics have also been used since they were invented; they can be molded with length markings instead of being scribed. Metal is used for more durable rulers for use in the workshop; sometimes a metal edge is embedded into a wooden desk ruler to preserve the edge when used for straight-line cutting. 12 inches or 30 cm in length is useful for a ruler to be kept on a desk to help in drawing.

Shorter rulers are convenient for keeping in a pocket. Longer rulers, e.g., 18 inches (45 cm) are necessary in some cases. Rigid wooden or plastic yardsticks, 1 yard long and metre sticks, 1 metre long, are also used. Classically, long measuring rods were used for larger projects, now superseded by tape measure or laser rangefinders.

Desk rulers are used for three main purposes: to measure, to aid in drawing straight lines and as a straight guide for cutting and scoring with a blade. Practical rulers have distance markings along their edges.

A type of ruler used in the printing industry is called a line gauge. These may be made from a variety of materials, typically metal or clear plastic. Units of measurement on a basic line gauge usually include inches, agate, picas, and points. More detailed line gauges may contain sample widths of lines, samples of common type in several point sizes, etc.

Measuring instruments similar in function to rulers are made portable by folding (carpenter's folding rule) or retracting into a coil (metal tape measure) when not in use.

When extended for use they are straight, like a ruler. The illustrations on this page show a 2-metre carpenter's rule which folds down to a length of 24 cm to easily fit in a pocket, and a 5-metre-long tape which retracts into a small housing.

A flexible length measuring instrument which is not necessarily straight in use is the tailor's fabric tape measure, a length of tape calibrated in inches and centimetres. It is used to measure around a solid body, e.g., a person's waist measurement, as well as linear measurement, e.g., inside leg. It is rolled up when not in use, taking up little space.

A *contraction rule* is made having larger divisions than standard measures to allow for shrinkage of a metal casting. They may also be known as a 'shrinkage *or* shrink *rule.*

A *ruler software programme* can be used to measure pixels on a computer screen.

Ruler Applications in Geometry

In geometry, a ruler without any marks on it (a straightedge) may be used only for drawing straight lines between points. A straightedge is also used to help draw accurate graphs and tables.

A ruler and compass construction refers to constructions using an unmarked ruler and a compass. It is possible to bisect an angle into two equal parts with ruler and compass. It can be proved, though, that it is impossible to divide an angle into three equal parts using only a compass and straightedge — the problem of angle trisection. However, should two marks be allowed on the ruler, the problem becomes solvable.

History

***Figure:** A wooden carpenter's rule and other tools found on board the 16th century carrack* Mary Rose

Rulers made of Ivory were in use by the Indus Valley Civilization period prior to 1500 BC. Excavations at Lothal (2400 BC) have yielded one such ruler calibrated to about ^{1}D $_{16}$ inches (1.6 millimetres). Ian Whitelaw holds that the Mohenjo-Daro ruler is divided into units corresponding to 1.32 inches (33.5 millimetres) and these are marked out in decimal subdivisions with amazing accuracy, to within 0.005 inches (0.13 millimetres). Ancient bricks found throughout the region have dimensions that correspond to these units. Anton Ullrich invented the folding ruler in 1851.

Curved and Flexible Rulers

The equivalent of a ruler for drawing or reproducing a smooth curve is known, where it takes the form of a rigid template, as a French curve. A flexible device which can be bent to the desired shape is known as a flat spline, or (in its more modern incarnation) a *flexible curve.* Historically, a flexible lead rule used by masons that could be bent to the curves of a molding was known as a lesbian rule.

Philosophy

Ludwig Wittgenstein famously used rulers as an example in his discussion of language games in the Philosophical Investigations. He pointed out that the standard metre bar in Paris was the criterion against which all other rulers were determined to be one metre long, but that there was no analytical way to demonstrate that the standard metre bar itself was one metre long. It could only be asserted as one metre as part of a language game.

Sticking to the Ruler's Marks

When using a ruler use the smallest mark as the first estimated digit. For example if a ruler's smallest mark is cm, and 4.5 cm is read, it is 4.5 (+-.1) or range 4.4 to 4.6 cm.

The overal length of a ruler may not be accurate within the degree of the smallest mark and the marks may be imperfectly spaced within each unit. With both of these errors in mind, reading between the marks (to 4.55) does not increase one's confidence in the overall reading and may make it worse.

Caliper

A caliper (British spelling also calliper, or in plurale tantum sense a pair of callipers) is a device used to measure the distance between two opposite sides of an object. A caliper can be as simple as a compass with inward or outward-facing points. The tips of the caliper are adjusted to fit across the points to be measured, the caliper is then removed and the distance read by measuring between the tips with a measuring tool, such as a ruler.

It is used in many fields such as mechanical engineering, metalworking, forestry, woodworking, science and medicine.

Nomenclature Variants

A plurale tantum sense of the word "calipers" coexists in natural usage with the regular noun sense of "caliper". That is, sometimes a caliper is treated cognitively like a pair of glasses or a pair of scissors, resulting in a phrase such as "hand me those calipers" or "those calipers are mine" in reference to one unit.

Also existing colloquially but not in formal usage is referring to a vernier caliper as a "vernier" or a "pair of verniers". In imprecise colloquial usage, some speakers extend this even to dial calipers, although they involve no vernier scale.

In machine-shop usage, the term "caliper" is often used in contradistinction to "micrometre", even though outside micrometres are technically a form of caliper. In this usage, "caliper" implies only the form factor of the vernier or dial caliper (or its digital counterpart).

History

The earliest caliper has been found in the Greek Giglio wreck near the Italian coast. The ship find dates to the 6th century BC. The wooden piece already featured a fixed and a movable jaw. Although rare finds, caliper remained in use by the Greeks and Romans.

A bronze caliper, dating from 9 AD, was used for minute measurements during the Chinese Xin dynasty. The caliper had an inscription stating that it was "made on a *gui-you* day at new moon of the first month of the first year of the Shijian guo period." The calipers included a "slot and pin" and "graduated in inches and tenths of an inch."

The modern vernier caliper, reading to thousandths of an inch, was invented by American Joseph R. Brown in 1851. It was the first practical tool for exact measurements that could be sold at a price within the reach of ordinary machinists.

Types

Inside Caliper

The inside calipers are used to measure the internal size of an object.

- The upper caliper in the image (at the right) requires manual adjustment prior to fitting, fine setting of this caliper type is performed by tapping the caliper legs lightly on a handy surface until they will *almost* pass over the object. A light push against the resistance of the central pivot screw then spreads the legs to the correct dimension and provides the required, consistent *feel* that ensures a repeatable measurement.

- The lower caliper in the image has an adjusting screw that permits it to be carefully adjusted without removal of the tool from the workpiece.

Outside Caliper

Outside calipers are used to measure the external size of an object.

The same observations and technique apply to this type of caliper, as for the above inside caliper. With some understanding of their limitations and usage these instruments can provide a high degree of accuracy and repeatability. They are especially useful when measuring over very large distances, consider if the calipers are used to measure a large diametre pipe. A vernier caliper does not have the depth capacity to straddle this large diametre while at the same time reach the outermost points of the pipe's diametre. They are made from high carbon steel.

Divider Caliper

In the metalworking field, a divider caliper, popularly called a compass, is used in the process of marking out locations. The points are sharpened so that they act as scribers, one leg can then be placed in the dimple created by a centre or prick punch and the other leg pivoted so that it scribes a line on the workpiece's surface, thus forming an arc or circle.

A divider caliper is also used to measure a distance between two points on a map. The two caliper's ends are brought to the two points whose distance is being measured. The caliper's opening is then either measured on a separate ruler and then converted to the actual distance, or it is measured directly on a scale drawn on the map. On a nautical chart the distance is often measured on the latitude scale appearing on the sides of the map: one minute of arc of latitude is approximately one nautical mile or 1852 metres.

Dividers are also used in the medical profession. An ECG (also EKG) caliper transfers distance on an electrocardiogram; in conjunction with the appropriate scale, the heart rate can be determined. A pocket caliper versions was invented by cardiologist Robert A. Mackin.

Oddleg Caliper

Ddleg calipers, Hermaphrodite calipers, or Oddleg jennys, as pictured on the left, are generally used to scribe a line a set distance from the edge of a workpiece. The bent leg is used to run along the workpiece edge while the scriber makes its mark at a predetermined distance, this ensures a line parallel to the edge. In the diagram at left, the uppermost caliper has a slight shoulder in the bent leg allowing it to sit on the edge more securely, the lower caliper lacks this feature but has a renewable scriber that can be adjusted for wear, as well as being replaced when excessively worn.

Vernier Caliper

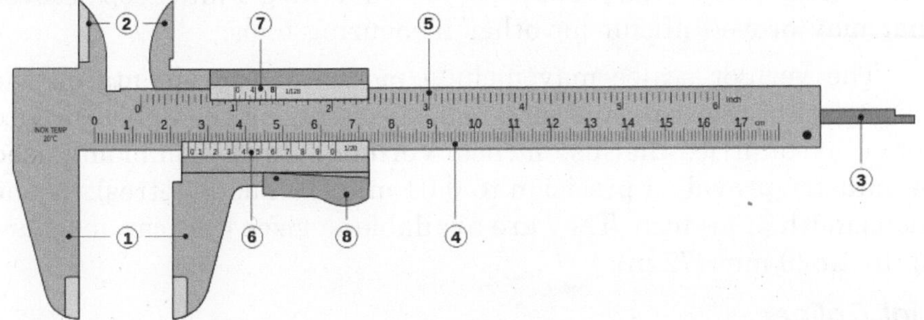

Parts of a vernier caliper:

1. Outside jaws: used to measure external diametre or width of an object
2. Inside jaws: used to measure internal diametre of an object
3. Depth probe: used to measure depths of an object or a hole
4. Main scale: scale marked every mm
5. Main scale: scale marked in inches and fractions
6. Vernier scale gives interpolated measurements to 0.1 mm or better
7. Vernier scale gives interpolated measurements in fractions of an inch
8. Retainer: used to block movable part to allow the easy transferring of a measurement

The vernier, dial, and digital calipers give a direct reading of the distance measured with high accuracy and precision. They are functionally identical, with different ways of reading the result. These calipers comprise a calibrated scale with a fixed jaw, and another jaw, with a pointer, that slides along the scale. The distance between the jaws is then read in different ways for the three types.

The simplest method is to read the position of the pointer directly on the scale. When the pointer is between two markings, the user can mentally interpolate to improve the precision of the reading. This would be a simple calibrated caliper; but the addition of a vernier scale allows more accurate interpolation, and is the universal practice; this is the vernier caliper.

Vernier, dial, and digital calipers can measure internal dimensions (using the uppermost jaws in the picture at right), external dimensions using the pictured lower jaws, and in many cases depth by the use of a probe that is attached to the movable head and slides along the centre of the body. This probe is slender and can get into deep grooves that may prove difficult for other measuring tools.

The vernier scales may include metric measurements on the lower part of the scale and inch measurements on the upper, or vice versa, in countries that use inches. Vernier calipers commonly used in industry provide a precision to 0.01 mm (10 micrometres), or one thousandth of an inch. They are available in sizes that can measure up to 1,829 mm (72 in).

Dial Caliper

Instead of using a vernier mechanism, which requires some practice to use, the dial caliper reads the final fraction of a millimetre or inch on a simple dial.

In this instrument, a small, precise gear rack drives a pointer on a circular dial, allowing direct reading without the need to read a vernier scale. Typically, the pointer rotates once every inch, tenth of an inch, or 1 millimetre. This measurement must be added to the coarse whole inches or centimetres read from the slide. The dial is usually arranged to be rotatable beneath the pointer, allowing for "differential" measurements (the measuring of the difference in size between two objects, or the setting of the dial using a master object and subsequently being able to read directly the plus-or-minus variance in size of subsequent objects relative to the master object).

The slide of a dial caliper can usually be locked at a setting using a small lever or screw; this allows simple go/no-go checks of part sizes.

Digital Caliper

A refinement now popular is the replacement of the analog dial with an electronic digital display on which the reading is displayed as a single value. Some digital calipers can be switched between centimetres or millimetres, and inches. All provide for zeroing the display at any point along the slide, allowing the same sort of differential measurements as with the dial caliper. Digital calipers may contain some sort of "reading hold" feature, allowing the reading of dimensions even in awkward locations where the display cannot be seen. Ordinary 6-in/150-mm digital calipers are made of stainless steel, have a rated accuracy of 0.001 in (0.02mm) and resolution of 0.0005 in (0.01 mm). The same technology is used to make longer 8-in and 12-in calipers; the accuracy for bigger measurements declines to 0.001 in (0.03 mm) for 100–200 mm and 0.0015 in (0.04 mm) for 200–300 mm.

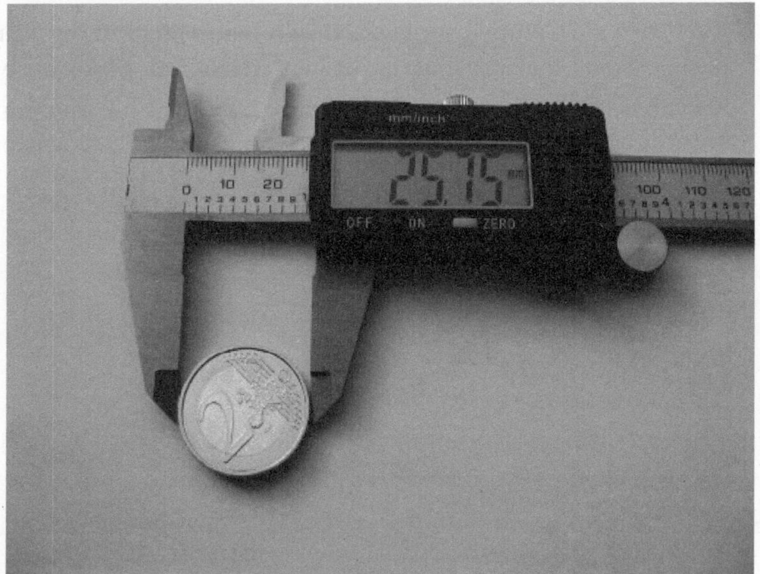

Figure: *Digital caliper*

Many Chinese-made digital calipers are inexpensive and perform reasonably well. One point worth noting is battery current when they are turned off. Many calipers do not stop drawing power when the switch is in the off position; they shut down the display but continue drawing nearly as much current. The current may be as much as 20 microamperes, which is much higher than many established brands. Sometimes calipers may not work properly when the battery voltage has dropped relatively little; silver cells, preferably selected from a datasheet to have a constant voltage for most of their life, may give a much longer usable life than alkaline button cells (e.g., SR44 instead of LR44).

Increasingly, digital calipers offer a serial data output to allow them to be interfaced with a dedicated recorder or a personal computer. The digital interface significantly decreases the time to make and record a series of measurements, and it also improves the reliability of the records. A suitable device to convert the serial data output to common computer interfaces such as RS-232, Universal Serial Bus, or wireless can be built or purchased. With such a converter, measurements can be directly entered into a spreadsheet, a statistical process control programme, or similar software.

The serial digital output varies among manufacturers. Common options are

- Mitutoyo's Digimatic interface. This is the dominant name brand interface. Format is 52-bits arranged as 13 nibbles.
- Sylvac interface. This is the common protocol for inexpensive, non-name brand, calipers. Format is 24 bit 90 kHz synchronous.
- Starrett
- Brown & Sharpe
- Federal
- Mahr (appears to offer Digimatic, RS232, and USB)
- Tesa
- Aldi. Format is 7 BCD digits.

Like dial calipers, the slide of a digital caliper can usually be locked using a lever or thumb-screw.

Some digital calipers contain a capacitive linear encoder. A pattern of bars is etched directly on the printed circuit board in the slider. Under the scale of the caliper another printed circuit board also contains an etched pattern of lines. The combination of these printed circuit boards forms two variable capacitors. The two capacitances are out of phase. As the slider moves the capacitance changes in a linear fashion and in a repeating pattern. The circuitry built into the slider counts the bars as the slider moves and does a linear interpolation based on the magnitudes of the capacitors to find the precise position of the slider. Other digital calipers contain an inductive linear encoder, which allows robust performance in the presence of contamination such as coolants. Magnetic linear encoders are used in yet other digital calipers.

Micrometre Caliper

A caliper using a calibrated screw for measurement, rather than a slide, is called a *micrometre caliper* or, more often, simply a *micrometre*. (Sometimes the term *caliper*, referring to any other type in this article, is held in contradistinction to *micrometre*.)

Comparison

Each of the above types of calipers have their relative merits and faults.

Vernier calipers are rugged and have long lasting accuracy, are coolant proof, are not affected by magnetic fields, and are largely shock proof. They may have both centimetre and inch scales. However, vernier calipers require good eyesight or a magnifying glass to read and can be difficult to read from a distance or from awkward angles. It is relatively easy to misread the last digit. In production environments, reading vernier calipers all day long is error-prone and is annoying to the workers.

Dial calipers are comparatively easy to read, especially when seeking exact centre by rocking and observing the needle movement. They can be set to 0 at any point for comparisons. They are usually fairly susceptible to shock damage. They are also very prone to getting dirt in the gears, which can cause accuracy problems.

Digital calipers switch easily between centimetre and inch systems.They can be set to 0 easily at any point with full count in either direction, and can take measurements even if the display is completely hidden, either by using a "hold" key, or by zeroing the display and closing the jaws, showing the correct measurement, but negative. They can be mechanically and electronically fragile. Most also require batteries, and do not resist coolant well. They are also only moderately shockproof, and can be vulnerable to dirt.

Calipers may read to a resolution of 0.01 mm or 0.0005 in, but accuracy may not be better than about ±0.02 mm or 0.001 in for 150 mm (6 in) calipers, and worse for longer ones.

Use

A caliper must be properly applied against the part in order to take the desired measurement. For example, when measuring the thickness of a plate a vernier caliper must be held at right angles to the piece. Some practice may be needed to measure round or irregular objects correctly.

Accuracy of measurement when using a caliper is highly dependent on the skill of the operator. Regardless of type, a caliper's jaws must be forced into contact with the part being measured. As both part and caliper are always to some extent elastic, the amount of force used affects the indication. A consistent, firm touch is correct. Too much force results in an underindication as part and tool distort; too little force gives insufficient contact and an overindication. This is a greater problem with a caliper incorporating a wheel, which lends mechanical advantage.

This is especially the case with digital calipers, calipers out of adjustment, or calipers with a poor quality beam.

Simple calipers are uncalibrated; the measurement taken must be compared against a scale. Whether the scale is part of the caliper or not, all analog calipers—verniers and dials—require good eyesight in order to achieve the highest precision. Digital calipers have the advantage in this area.

Calibrated calipers may be mishandled, leading to loss of zero. When a calipers' jaws are fully closed, it should of course indicate zero. If it does not, it must be recalibrated or repaired. It might seem that a vernier caliper cannot get out of calibration but a drop or knock can be enough. Digital calipers have zero set buttons.

Vernier, dial and digital calipers can be used with accessories that extend their usefulness. Examples are a base that extends their usefulness as a depth gauge and a jaw attachment that allows measuring the centre distance between holes. Since the 1970s a clever modification of the moveable jaw on the back side of any caliper allows for step or depth measurements in addition to external caliper measurements, in similar fashion to a universal micrometre (e.g., Starrett Mul-T-Anvil or Mitutoyo Uni-Mike).

Zero Error

The method to use a vernier scale or caliper with zero error is to use the formula 'actual reading = main scale + vernier scale " (zero error)'. Zero error may arise due to knocks that cause the calibration at the 0.00 mm when the jaws are perfectly closed or just touching each other.

Positive zero error refers to the fact that when the jaws of the vernier caliper are just closed, the reading is a positive reading away from the actual reading of 0.00mm. If the reading is 0.10mm, the zero error is referred to as +0.10mm.

Negative zero error refers to the fact that when the jaws of the vernier caliper are just closed, the reading is a negative reading away from the actual reading of 0.00mm. If the reading is 0.08mm, the zero error is referred to as –0.08mm.

Engineering Drawing And Workshop Drawing

Engineering drawing is a language which is understood throughout the world by engineers and fabricators. Other languages may fail to describe the size, shape, physical aspects, inner details, finish, etc., but the engineers' language known as '*Engineering Drawing*' never fails. The most intricate assemblies with their various complicated parts can be easily represented by engineering graphics. There are various types of engineering drawings and all have one simple purpose, i.e. the communication of ideas to others. The drawings used for communicating the ideas of the design engineer to the production engineer and technicians is known as '*Workshop Drawing*'.

Engineering Drawing

A drawing Prepared by an engineer, for an engineering purpose is known as an engineering drawing. It is the graphic representation of physical objects and their relationship. It is prepared, based on certain basic principles, symbolic representations, standard conventions, notations, etc. It is the only universal means of communication used by engineers and technicians.

Pictorial Drawing

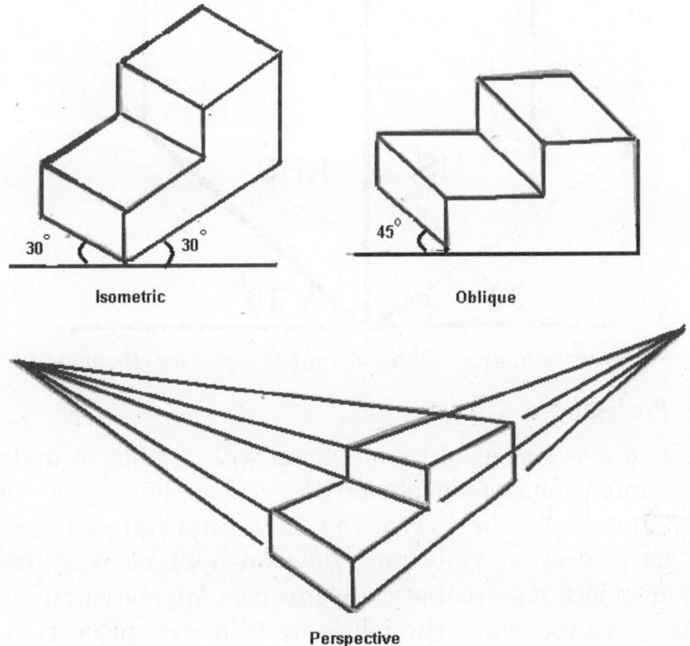

Every person cannot understand the orthographic projection. Its execution requires a thorough understanding of the principles of projection and its reading requires a good practice of constructive imagination. We can describe the shape of a job by means of pictorial drawing also, which can be understood quite easily. Pictorial drawing is the drawing of a picture in graphic language of engineers, to represent a real thing by means of picture views. It shows the appearance of the object by one view only. Following three methods of pictorial projections are commonly used in engineering drawing:

(a) Isometric projection

(b) Oblique projection

(c) Perspective projection

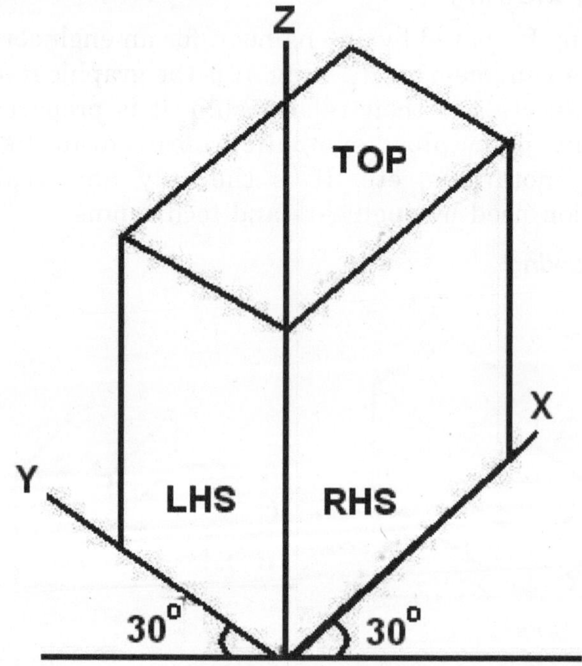

Figure: *Isometric, Oblique and Perspective Drawing*

Sometric Projection

A trained eye and good imagination will be able to understand the three dimensions of an object. Several orthographic views on different planes are to be drawn to understand fully an object. But in isometric projection, only one view on a plane is sufficient to represent an object in its realistic appearance. Anyone can understand by looking at a view what the job is by isometric projection.

Isometric projection is a type of pictorial projection. Isometric means equal measure. In this isometric projection, all the plane surfaces and the edges formed of these plane surfaces should be equally inclined to the metric plane. Metric plane is the same horizontal plane which is used in orthographic projection.

To represent the three dimensions (length, breadth and height) of the object, there are three axes known as ISOMETRIC AXES. To start an isometric drawing, a reference line (horizontal line) and the three axes (X, Y Z) are drawn by taking an angle of 30° from the reference horizontal line as drawn in figure Z axis is a vertical line to the horizontal line drawn from intersection point of X and Y axes.

Oblique Projection.

Oblique projection may be illustrated in different ways, according to the choice of axes, length of inclined side and direction of looking the side. As compared with isometric projection, in oblique projection, one side of the object is horizontal, second side is vertical and the third side is inclined at 30°or 45° to the horizontal. The lengths of the horizontal and vertical sides are equal to the actual lengths, but the length of the inclined side is taken as three-fourth or half of the actual length.

In oblique projection, an object is placed with its front face parallel to a vertical plane of projection and the visual rays parallel to each other pierce the plane of projection obliquely (Oblique means inclined). The projection represents the front face of the object in its true shape and size. The rest of the object is not projected true in its shape and size. Both the isometric and oblique projections are the methods of representing the object pictorially. But the oblique projection is preferable to the isometric projection in representing the objects of circular shapes, because the front face is in actual shape and size. It is not distorted.

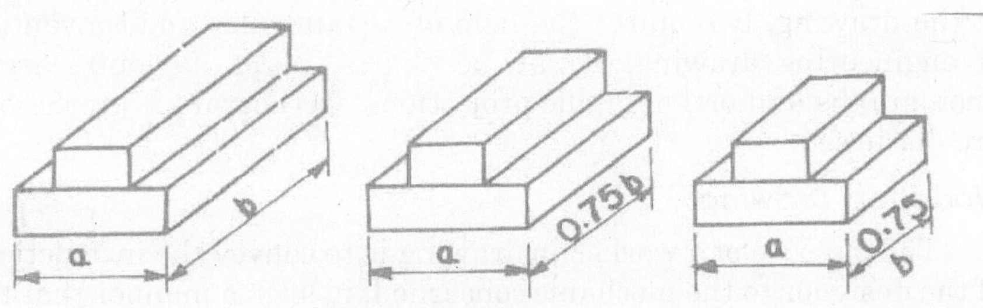

Figure: *Oblique Projection*

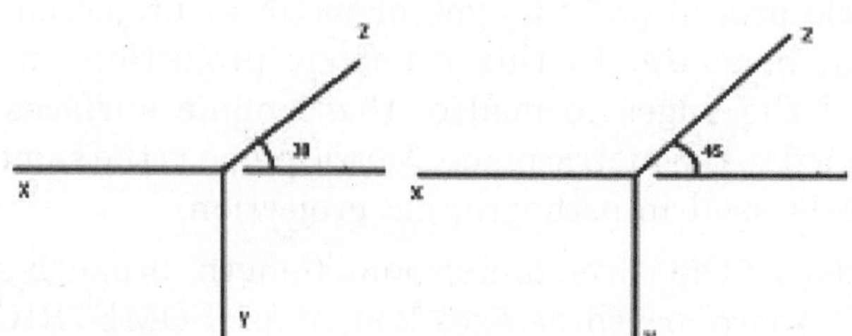

Figure: *Third Axis of the Oblique Projection may be Inclined at 30° or 45°*

Perspective Projection.

Perspective projection or perspective drawing is the representation of an object on a plane surface, called the picture plane, as it would appear to the eye, when viewed from a fixed position.

It may also be defined as the figure formed on the picture plane when visual rays from the eye to the object cut the picture plane. Perspective is mainly used in architecture. By means of perspective, the architecture is able to show how an object would appear when constructed.

Figure: *Perspective Projection*

Non-Pictorial Drawing

It is not always possible to visualize the object merely by looking at the drawing. It requires the help of certain rules and convention of engineering drawing. It can be charts, maps, lay-outs, aerial photographs and orthographic projections. Orthographic projections are discussed.

Workshop Drawing

The purpose of a workshop drawing is to convey the instructions of the designer to the mechanic concerned, in such a manner that the work to be done can be completed with accuracy and as rapidly as possible. The drawing will show all necessary measurements, which are expressed clearly without the need for any calculation. The limits of accuracy or the class of fit and the raw materials from which the parts are to be made, are also indicated.

The drawing may not be of actual size, but for convenience may be made larger or smaller than the actual job, i.e. it will be drawn to scale. The scale is indicated in the title block. Owing to usage and the possibility of being torn, it is undesirable that the original drawing should be used in the workshop. Instead, a number of prints are reproduced.

Types of Workshop Drawing

There are three types of workshop drawing:

(a) *Detail Drawing:* This type of drawing provides all necessary dimensions and working instructions for the production of an item, e.g. one part of a hydraulic component.

(b) *Assembly Drawing*: This type of drawing shows the component or assembly in an assembled state, e.g. the assembly of the various parts of a hydraulic component.

(c) *General Arrangement Drawing:* This type of drawing shows a number of components in an assembled condition to form a self- contained unit, e.g. the complete hydraulic system of an aircraft.

8

Basic Engineering Drawing: Dimensioning and Layout Procedure

Introduction

A drawing of an object is prepared to define its shape and to specify its size. The shape description is based on projection and the size description on dimensioning. Every drawing must give its complete size description stating length, width, thickness, diametre of holes, grooves, angles, etc. and such other details relating to its construction. To give all those measurements and information describing the size of the object in the drawing is called dimensioning.

Placing of Dimensions

Dimensions should be placed on the view, which shows the relevant features most clearly. The two recommended systems of placing the dimensions are:

(a) *Aligned System:* In this system, all dimensions are so placed that these may be read from the bottom or the right hand edge of the drawing sheet. All dimensions should be placed above the dimension lines.

(b) *Unidirectional System:* In this system, all dimensions are so placed that these may be read from the bottom edge of the drawing sheet. In this system, there is no restriction controlling the direction of dimension lines. This system is advantageous on large drawings, where it is inconvenient to read dimensions from the right hand side. In this method, all dimension lines are interrupted, preferably near the middle for the insertion of the dimension value.

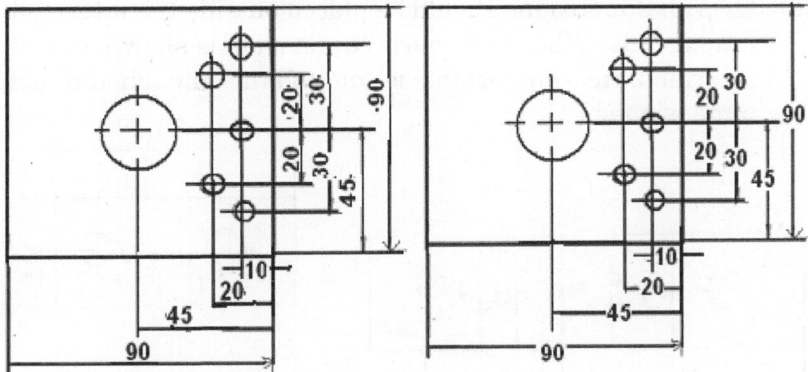

***Figuge:** Aligned System and Unidirectional System*

- General Principles of Dimensioning.

(a) As far as possible, all the dimensions for one particular operation shall be specified in one view only, such as diametre and depth of drilled hole, or size and depth of a threaded hole, etc.

(b) Normally dimensions should be placed outside the views, but if it is not possible it may be placed within the view as shown. However , dimensions should not be placed within a view unless drawing becomes clear by doing so. Dimensions should not be placed too close to each other or to the parts being dimensioned.

(c) Dimensions are to be given from visible outlines, rather than from hidden lines, Dimensions are to be given from a base line, a centre line of the hole, a cylindrical part, important hole or a finished surface which may be readily established, based on design requirements and the relationship to other parts.

(d) Dimensions for different operations on a part, for example, drilling and bending, should be given separately as in figure, if permissible by its design.

(e) An axis or a contour line should never be used as a dimension line but may be used as a projection line.

(f) The intersection of dimension line should be avoided as far as possible if, however, the intersection of two dimension lines is unavoidable, the lines should not be broken. Dimension lines may be broken for inserting the dimension in the case of unidirectional dimensioning.

(g) Overall dimensions should be placed outside the intermediate dimensions. Where an overall dimension is shown, one of the intermediate dimensions is redundant and should not be dimensioned.

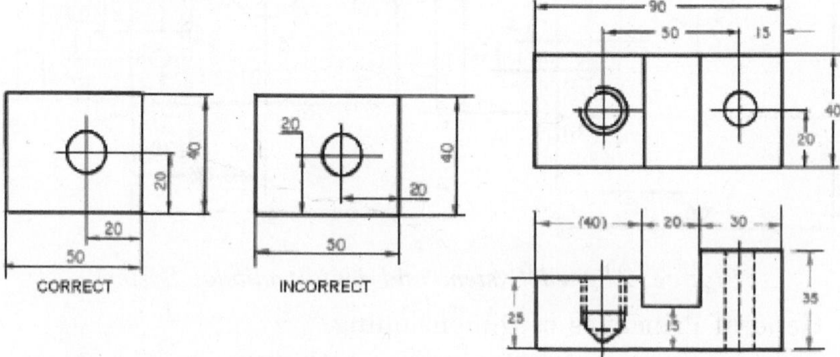

Figure: *Dimensions Placed Outside the View and Dimensions Placed within the View*

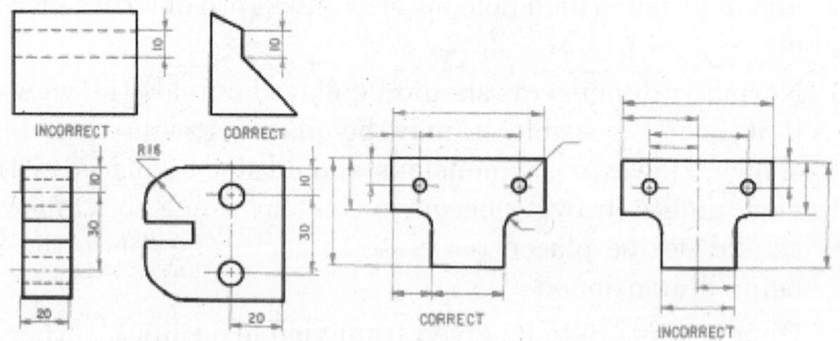

Figure: *Dimensions for Visible Outline and Dimensions to a Centre Line*

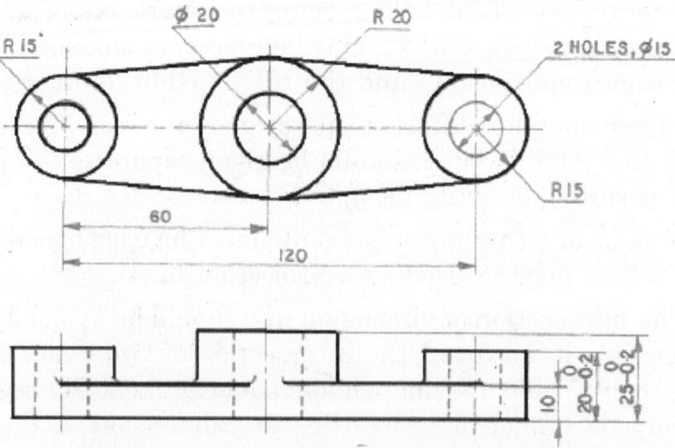

Figure: *Dimensioning the Cylindrical Parts*

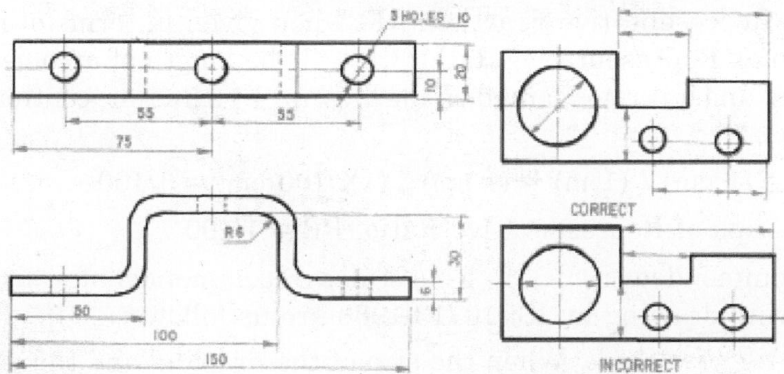

***Figure:** Dimensions for Different Operations used as Dimension Line and Axis or Contour Line not.*

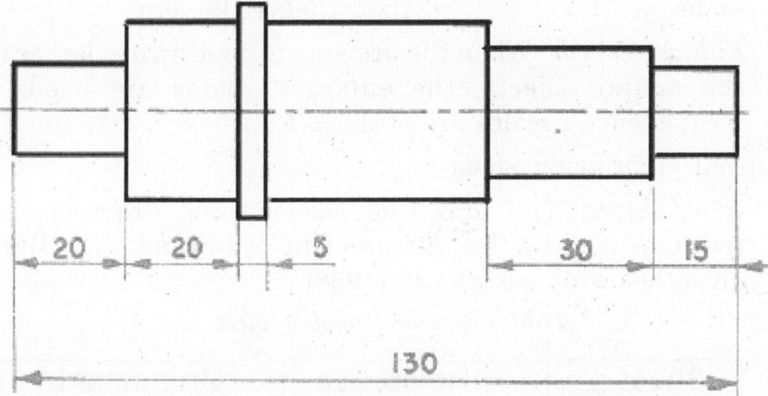

***Figure:** Overall Dimension Outside the Intermediate Dimensions*

Scales

Drawing of very big objects cannot be prepared in full size because these would be too big to accommodate on the drawing sheet. Drawings of very small objects also cannot be prepared in full size because these would be too small to draw and to read. A convenient scale is chosen to prepare the drawings of big as well as small objects in proportionately smaller or larger size. Therefore, scales are used to prepare a drawing at a full size, reduced size or enlarged size.

Representative Fraction. The ratio of the size of the drawing to the size of the object is known as the representative fraction. It is denoted as RF.

$$\text{Representative Fraction (RF)} = \frac{\text{Dimension on drawing}}{\text{Actual dimension}}$$

The Representative Fraction (RF) when given in terms of ratio is known as Representative Ratio (RR). If the length of an object is one metre and it is represented on the drawing by a line one centimetre long, then

RF = (1 cm) / (1 m) = (1 cm) / (1X 100 cm) = 1/100

In terms of Representative Ratio, RR = 1:100

Recommended Scales. The scales recommended for use in engineering drawing by IS: 10713-1983 are as follows:

(a). *Full Size Scale:* When the size of the drawing and the object is the same, then it is known as full-size scale, i.e. 1:1.

(b) *Reduced Scale:* When the drawings are smaller in size than the actual objects, the reduced scales are used. Recommended scales are 1:2, 1:5, 1:10, 1:20, 1:50, 1:100, etc.

(c) *Enlarged Scale:* When the drawing to be drawn is larger than the actual objects, the enlarged scales are used. The recommended scales are 50:1, 20:1, 10:1, 5:1, 2:1, etc.

Layout of Drawing Sheet.

(a) *Sheet Sizes:* The preferred sizes of the drawing sheets recommended by the Bureau of Indian Standards (BIS) are given below as per SP: 46 (1988).

Table: *Size of Drawing sheet*

Sheet Designation	Trimmed size (mm)	Untrimmed size(mm)
A0	841 X 1189	880X1230
A1	594X841	625X880
A2	420X594	450X625
A3	297X420	330X450
A4	210X297	240X330
A5	147X210	165X240

The layout of the drawing on a drawing sheet should be done in such a manner so as to make its reading easy. All dimensions are in millimetres. A full *size* drawing paper is normally of 565 mm x 765 mm size.

(b) *Margin:* Margin is provided in the drawing sheet by drawing margin lines. Prints are trimmed along these lines. After trimming, the prints would be of the recommended sizes of the trimmed sheets.

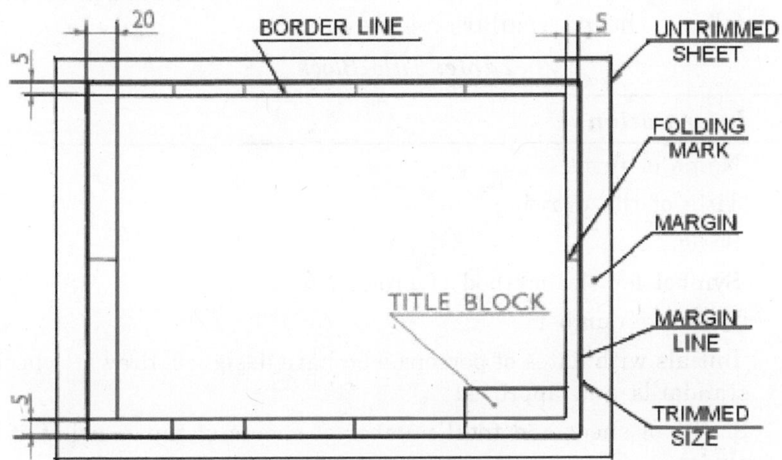

Figure: *Margin*

(c) *Border Lines:* Clear working space is obtained by drawing border *lines* as shown. More space is kept on the left-hand side for the purpose of filing or binding if necessary. When prints are to be preserved or stored in a cabinet without filing, equal space may be provided on all sides.

(d) *Borders and Frames.* SP: 46 (1988) recommends the borders of 20 mm width *for* the sheet sizes A0 and A1 and 10 mm for the sizes A2 to A5. Frame shows the clear space available for the drawing purpose.

(e) *Orientation Mark:* Four centering marks are drawn as shown in figure to facilitate positioning of the drawing for the reproduction purpose. The *orientation mark* will coincide with one of the centering marks which can be used for the orientation of drawing sheet on the drawing board.

(a) *Grid Reference System* (Zone system): The *grid reference system* is drawn on the sheet to permit easy location on the drawing such as details, alterations or additions. The rectangle of grid along the length should be referred by numerals 1, 2, 3... and along the width by the capital letters A, B, C, D, etc. as shown in figure.

(b) *Title Block:* Space for the title block must be provided in the bottom right-hand corner of the drawing sheet as shown in figure. The size of the title block as recommended by the BIS is 185 mm x 65 mm for all designations of the drawing sheets. The simplest type of a title block. All title blocks should contain at least the particulars as shown

***Table:** Title Block*

Sl No.	*Information*
1	Name of firm
2	Title of the drawing
3	Scale
4	Symbol for the method of projection
5	Drawing number
6	Initials with dates of persons who have designed, drawn, checked standards and approved.
7	Sl No. of sheet and total number of sheets of the drawing of the object

Computer-aided Design

Computer-aided design (CAD) is the use of computer systems to assist in the creation, modification, analysis, or optimization of a design. CAD software is used to increase the productivity of the designer, improve the quality of design, improve communications through documentation, and to create a database for manufacturing. CAD output is often in the form of electronic files for print, machining, or other manufacturing operations.

Computer-aided design is used in many fields. Its use in electronic design is known as Electronic Design Automation, or EDA. In mechanical design is known as Mechanical Design Automation, or MDA, it is also known as computer-aided drafting (CAD) which describes the process of creating a technical drawing with the use of computer software.

CAD software for mechanical design uses either vector based graphics to depict the objects of traditional drafting, or may also produce raster graphics showing the overall appearance of designed objects. However, it involves more than just shapes. As in the manual drafting of technical and engineering drawings, the output of CAD must convey information, such as materials, processes, dimensions, and tolerances, according to application-specific conventions.

CAD may be used to design curves and figures in two-dimensional (2D) space; or curves, surfaces, and solids in three-dimensional (3D) space.

CAD is an important industrial art extensively used in many applications, including automotive, shipbuilding, and aerospace industries, industrial and architectural design, prosthetics, and many more. CAD is also widely used to produce computer animation for special effects in movies, advertising and technical manuals, often called DCC Digital content creation. The modern ubiquity and power of computers means that even perfume bottles and shampoo dispensers are designed using techniques unheard of by engineers of the 1960s. Because of its enormous economic importance, CAD has been a major driving force for research in computational geometry, computer graphics (both hardware and software), and discrete differential geometry.

The design of geometric models for object shapes, in particular, is occasionally called *computer-aided geometric design* (*CAGD*).

While the goal of automated CAD systems is to increase efficiency, they are not necessarily the best way to allow newcomers to understand the geometrical principles of Solid Modelling. For this, scripting languages such as PLaSM (Programming Language of Solid Modelling) are more suitable.

Overview

Beginning in the 1980s computer-aided design programmes reduced the need of draftsmen significantly, especially in small to mid-sized companies. Their affordability and ability to run on personal computers also allowed engineers to do their own drafting and analytic work, eliminating the need for entire departments. In today's world, many students in universities do not learn manual drafting techniques because they are not required to do so. The days of hand drawing for final drawings are virtually over. Universities no longer require the use of protractors and compasses to create drawings, instead there are several classes that focus on the use of CAD software.

Current computer-aided design software packages range from 2D vector-based drafting systems to 3D solid and surface modelers. Modern CAD packages can also frequently allow rotations in three dimensions, allowing viewing of a designed object from any desired angle, even from the inside looking out. Some CAD software is capable of dynamic mathematical modelling, in which case it may be marketed as CADD.

CAD is used in the design of tools and machinery and in the drafting and design of all types of buildings, from small residential types (houses) to the largest commercial and industrial structures (hospitals and factories).

CAD is mainly used for detailed engineering of 3D models and/ or 2D drawings of physical components, but it is also used throughout the engineering process from conceptual design and layout of products, through strength and dynamic analysis of assemblies to definition of manufacturing methods of components. It can also be used to design objects. Furthermore many CAD applications now offer advanced rendering and animation capabilities so engineers can better visualize their product designs.4D BIM is a type of virtual construction engineering simulation incorporating time or schedule related information for project management.

CAD has become an especially important technology within the scope of computer-aided technologies, with benefits such as lower product development costs and a greatly shortened design cycle. CAD enables designers to layout and develop work on screen, print it out and save it for future editing, saving time on their drawings.

Uses

Computer-aided design is one of the many tools used by engineers and designers and is used in many ways depending on the profession of the user and the type of software in question.

CAD is one part of the whole Digital Product Development (DPD) activity within the Product Lifecycle Management (PLM) processes, and as such is used together with other tools, which are either integrated modules or stand-alone products, such as:

- Computer-aided engineering (CAE) and Finite element analysis (FEA)
- Computer-aided manufacturing (CAM) including instructions to Computer Numerical Control (CNC) machines
- Photo realistic rendering
- Document management and revision control using Product Data Management (PDM).

CAD is also used for the accurate creation of photo simulations that are often required in the preparation of Environmental Impact Reports, in which computer-aided designs of intended buildings are superimposed into photographs of existing environments to represent what that locale will be like were the proposed facilities allowed to be built. Potential blockage of view corridors and shadow studies are also frequently analyzed through the use of CAD.

CAD has been proven to be useful to engineers as well. Using four properties which are history, features, parametreization, and high level constraints. The construction history can be used to look back into the model's personal features and work on the single area rather than the whole model. Parametres and constraints can be used to determine the size, shape, and other properties of the different modelling elements. The features in the CAD system can be used for the variety of tools for measurement such as tensile strength, yield strength, electrical or electro-magnetic properties. Also its stress, strain, timing or how the element gets affected in certain temperatures, etc.

Types

There are several different types of CAD, each requiring the operator to think differently about how to use them and design their virtual components in a different manner for each.

There are many producers of the lower-end 2D systems, including a number of free and open source programmes. These provide an approach to the drawing process without all the fuss over scale and placement on the drawing sheet that accompanied hand drafting, since these can be adjusted as required during the creation of the final draft.

3D wireframe is basically an extension of 2D drafting (not often used today). Each line has to be manually inserted into the drawing. The final product has no mass properties associated with it and cannot have features directly added to it, such as holes. The operator approaches these in a similar fashion to the 2D systems, although many 3D systems allow using the wireframe model to make the final engineering drawing views.

3D "dumb" solids are created in a way analogous to manipulations of real world objects (not often used today). Basic three-dimensional geometric forms (prisms, cylinders, spheres, and so on) have solid volumes added or subtracted from them, as if assembling or cutting real-world objects. Two-dimensional projected views can easily be generated from the models. Basic 3D solids don't usually include tools to easily allow motion of components, set limits to their motion, or identify interference between components.

3D parametric solid modelling requires the operator to use what is referred to as "design intent". The objects and features created are adjustable. Any future modifications will be simple, difficult, or nearly impossible, depending on how the original part was created. One must think of this as being a "perfect world" representation of the component. If a feature was intended to be located from the centre of the part, the operator needs to locate it from the centre of the model, not, perhaps, from a more convenient edge or an arbitrary point, as he could when using "dumb" solids. Parametric solids require the operator to consider the consequences of his actions carefully.

Some software packages provide the ability to edit parametric and non-parametric geometry without the need to understand or undo the design intent history of the geometry by use of direct modelling functionality. This ability may also include the additional ability to infer the correct relationships between selected geometry (e.g., tangency, concentricity) which makes the editing process less time and labour intensive while still freeing the engineer from the burden of understanding the models. These kind of non-history based systems are called Explicit Modellers or Direct CAD Modelers.

Top end systems offer the capabilities to incorporate more organic, aesthetics and ergonomic features into designs. Freeform surface modelling is often combined with solids to allow the designer to create products that fit the human form and visual requirements as well as they interface with the machine.

Technology

Originally software for Computer-Aided Design systems was developed with computer languages such as Fortran, ALGOL but with the advancement of object-oriented programming methods this has radically changed. Typical modern parametric feature based modeler and freeform surface systems are built around a number of key C modules with their own APIs. A CAD system can be seen as built up from the interaction of a graphical user interface (GUI) with NURBS geometry and/or boundary representation (B-rep) data via a geometric modelling kernel. A geometry constraint engine may also be employed to manage the associative relationships between geometry, such as wireframe geometry in a sketch or components in an assembly.

Unexpected capabilities of these associative relationships have led to a new form of prototyping called digital prototyping. In contrast to physical prototypes, which entail manufacturing time in the design.

That said, CAD models can be generated by a computer after the physical prototype has been scanned using an industrial CT scanning machine. Depending on the nature of the business, digital or physical prototypes can be initially chosen according to specific needs.

Today, CAD systems exist for all the major platforms (Windows, Linux, UNIX and Mac OS X); some packages even support multiple platforms.

Right now, no special hardware is required for most CAD software. However, some CAD systems can do graphically and computationally intensive tasks, so a modern graphics card, high speed (and possibly multiple) CPUs and large amounts of RAM may be recommended.

The human-machine interface is generally via a computer mouse but can also be via a pen and digitizing graphics tablet. Manipulation of the view of the model on the screen is also sometimes done with the use of a Spacemouse/SpaceBall. Some systems also support stereoscopic glasses for viewing the 3D model.Technologies which in the past were limited to larger installations or specialist applications have become available to a wide group of users.These include the CAVE or HMD's and interactive devices like motion-sensing technology

History

Designers have long used computers for their calculations. Digital computers were used in power system analysis or optimization as early as proto-"Whirlwind" in 1949. Circuit design theory, or power network methodology would be algebraic, symbolic, and often Vector based. Examples of problems being solved in the mid-1940s to 50s include, Servo motors controlled by generated pulse (1949), The digital computer with built-in compute operations automatic co-ordinate transform to compute radar related vectors (1951) and the essentially graphic mathematical process of forming a shape with a digital machine tool (1952) were accomplished with the use of computer software. The man credited with coining the term CAD. Douglas T. Ross stated "As soon as I saw the interactive display equipment, [being used by radar operators 1953] I said "Gee, that's just what we need"". The designers of these very early computers built utility programmes so that programmers could debug programmes using flow charts on a display scope with logical switches that could be opened and closed during the debugging session. They found that they could create electronic symbols and geometric figures to be used to create simple circuit diagrams and flow charts. They made the pleasant discovery that an object once drawn could be reproduced at will, its orientation, Linkage [flux, mechanical, lexical scoping] or scale changed. This suggested numerous possibilities to them. It took ten years of interdisciplinary development work before SKETCHPAD sitting on evolving math librarys emerged from MIT`s labs. Additional developments were carried out in the 1960s within the aircraft, automotive, industrial control and electronics industries in the area of 3D surface construction, NC programming and design analysis, most of it independent of one another and often not publicly published until much later. Some of the mathematical description work on curves was developed in the early 1940s by Robert Issac Newton from Pawtucket, Rhode Island. Robert A. Heinlein in his 1957 novel *The Door into Summer* suggested the possibility of a robotic *Drafting Dan.* However, probably the most important work on polynomial curves and sculptured surface was done by Pierre Bézier (Renault), Paul de Casteljau (Citroen), Steven Anson Coons (MIT, Ford), James Ferguson (Boeing), Carl de Boor (GM), Birkhoff (GM) and Garibedian (GM) in the 1960s and W. Gordon (GM) and R. Riesenfeld in the 1970s.

It is argued that a turning point was the development of the SKETCHPAD system at MIT by Ivan Sutherland (who later created a graphics technology company with Dr. David Evans). The distinctive feature of SKETCHPAD was that it allowed the designer to interact with his computer graphically: the design can be fed into the computer by drawing on a CRT monitor with a light pen. Effectively, it was a prototype of graphical user interface, an indispensable feature of modern CAD. Sutherland presented his paper Sketchpad: A Man-Machine Graphical Communication System in 1963 at a Joint Computer Conference having worked on it as his PhD thesis paper for a few years. Quoting,"For drawings where motion of the drawing, or analysis of a drawn problem is of value to the user, Sketchpad excels. For highly repetitive drawings or drawings where accuracy is required, Sketchpad is sufficiently faster than conventional techniques to be worthwhile. For drawings which merely communicate with shops, it is probably better to use conventional paper and pencil." Over time efforts would be directed towards the goal of having the designers drawings communicate not just with shops but with the shop tool itself. This goal would be a long time arriving.

The first commercial applications of CAD were in large companies in the automotive and aerospace industries, as well as in electronics. Only large corporations could afford the computers capable of performing the calculations. Notable company projects were at GM (Dr. Patrick J.Hanratty) with DAC-1 (Design Augmented by Computer) 1964; Lockheed projects; Bell GRAPHIC 1 and at Renault (Bézier) – UNISURF 1971 car body design and tooling.

One of the most influential events in the development of CAD was the founding of MCS (Manufacturing and Consulting Services Inc.) in 1971 by Dr. P. J. Hanratty, who wrote the system ADAM (Automated Drafting And Machining) but more importantly supplied code to companies such as McDonnell Douglas (Unigraphics), Computervision (CADDS), Calma, Gerber, Autotrol and Control Data.

As computers became more affordable, the application areas have gradually expanded. The development of CAD software for personal desktop computers was the impetus for almost universal application in all areas of construction.

Other key points in the 1960s and 1970s would be the foundation of CAD systems United Computing, Intergraph, IBM, Intergraph IGDS in 1974 (which led to Bentley Systems MicroStation in 1984)

CAD implementations have evolved dramatically since then. Initially, with 3D in the 1970s, it was typically limited to producing drawings similar to hand-drafted drawings.

Advances in programming and computer hardware, notably solid modelling in the 1980s, have allowed more versatile applications of computers in design activities.

Key products for 1981 were the solid modelling packages -Romulus (ShapeData) and Uni-Solid (Unigraphics) based on PADL-2 and the release of the surface modeler CATIA (Dassault Systemes). Autodesk was founded 1982 by John Walker, which led to the 2D system AutoCAD.

The next milestone was the release of Pro/ENGINEER in 1988, which heralded greater usage of feature-based modelling methods and parametric linking of the parametres of features. Also of importance to the development of CAD was the development of the B-rep solid modelling kernels (engines for manipulating geometrically and topologically consistent 3D objects) Parasolid (ShapeData) and ACIS (Spatial Technology Inc.) at the end of the 1980s and beginning of the 1990s, both inspired by the work of Ian Braid. This led to the release of mid-range packages such as SolidWorks and TriSpective (later known as IronCAD) in 1995, Solid Edge (then Intergraph) in 1996 and Autodesk Inventor in 1999.

Engineering Drawing: An Historical Overview

Introduction

Just like literature uses language to interpret ideas and emotion, engineers use drawing to illustrate their ideas on paper. Furthermore, engineering drawing employs different tools in order to come up with meaningful sketches. These tools include lines, symbols, and conventions which bear different specific meanings to the engineers. Through these tools, an engineer cannot only draw meaningful sketches but also interpret the drawings of other engineers. There are generally two broad categories of engineering drawing; two dimensional drawing which is also referred to as plane geometry and three dimensional drawing also called solid geometry. Therefore, engineering drawing can be defined as a document that describes a precise object (Gupta, 2008).

History of engineering drawing

Development of engineering drawing was mainly influenced by changes in other industries such as mining, agriculture, transport and manufacturing especially between the 18th and 19th centuries. Pre-historically, human made sketches on the walls of rocks and caves mainly for communication purposes and for record keeping. Other purposes of these sketches were for religious and decorative reasons. These early sketches depicted pictures of animals and humans, and these early forms of drawing are believed to have laid the foundation for the development of engineering drawing. However, in the 14th and 15th centuries, engineering drawing seemed to have started to take shape since drawing representing building and machinery were widely used during this era.

They were normally pictures accompanied by detailed notes describing different aspects of the pictures. These drawing were referred to by craftsmen as they performed the actual construction of the objects represented in the drawing. However, these early drawing were devoid of specific dimension and the craftsmen used their discretion and experience to build the objects represented in the drawings. Over the centuries however, with the advent of computers, engineering drawing has been greatly enhanced through special computer software used to quickly and efficiently come up with engineering drawings. This software is generally referred to as Computer-Aided Design (CAD). Moreover, software has made engineering drawing fast and more accurate (Madsen & Madsen, 2011).

Major personalities that shaped engineering drawing

Engineering drawing was largely developed by artists such as Leonardo da Vinci. Some of his major artworks include Mona Lisa and The Last Supper which received enormous attention in Italy during the 15th century. These artists were called upon from time-to-time to draw sketches for many church building and other infrastructural works. Other than these drawings, da Vinci was a great inventor of machinery and one of his most renowned inventions included the glider machine and even warfare paraphernalia. Another great contributor to engineering drawing was Leon Battista Alberti since he was the first to incorporate geometry into his architectural works. Moreover, he specialized in town planning works and also ventured into perspective drawing to enhance the pictorial drawing that was the only known form of architectural drawing at the time (Shah & Rana, 2009). The Frenchman Rene Descartes in his contribution further insisted on the need for perspective drawing in order to represent more details in an architectural drawing. In order to attain this end, Descartes was the first artist to use the Cartesian plane for drawings and also discovered the analytic geometry that set the precedence for more accurate and detail drawings, a system that is widely utilized by CAD software.

Another great personality who contributed to development in engineering drawing was Gaspard Monge. He made his mark in the drawing arena by developing a large scale city plan using his own observation skills. He used what is now referred to as descriptive geometry in the planning of the city. In the late 18th century, engineers had developed modalities for developing geometrically accurate drawings and manufacturers and builder depended entirely on these sketches to manufacture their products (Gupta, 2008). David Michelangelo was also a great artist and sculptor whose architectural drawings were greatly used to build many churches and other prestigious building in the Baroque era and these building stood the test of time and are still standing strong up to present times. Some of them have been protected as monumental sites due to their design and aesthetics. These artists and inventors played critical roles in the development of engineering drawing as we know it today. Moreover, their sketches resulted in magnificent building especially during the Renaissance period in Italy, France and Germany and were highly valued by the Catholic Church for their talent and devotion (Shah & Rana, 2009).

Conclusion

Engineering drawing has evolved over the centuries from sketches of animals and humans on the walls of caves to intricate drawing of machine parts and buildings. This paradigm shift was largely triggered by exponential changes in other industries such as mining, agriculture and manufacturing. Contemporarily, development in information technology and computers in general has greatly simplified the process of engineering drawing through use of specialized software generally referred to as Computer-Aided Design (CAD) reducing the time and effort used for developing engineering drawing while at the same time developing more accurate drawings.

Helix

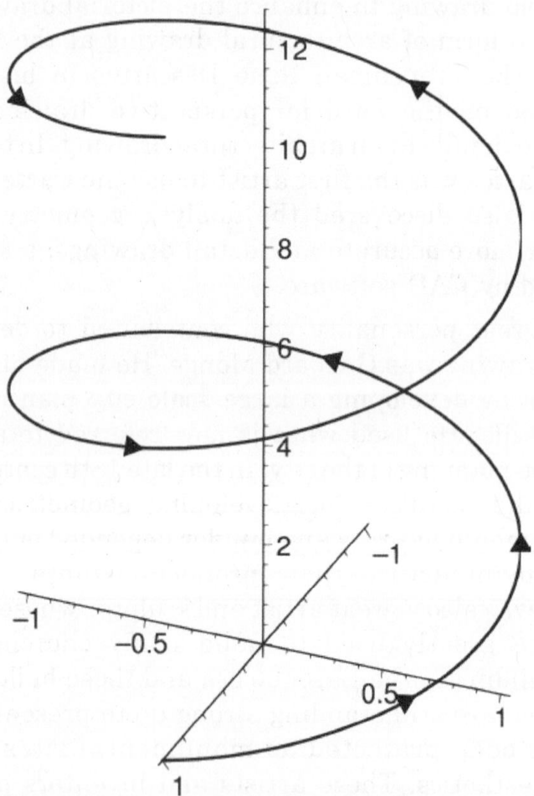

Figure: *The helix (cos* t*, sin* t*,* t*) from* t = *0 to 4π with arrowheads showing direction of increasing* t

A helix (pl: helixes or helices) is a type of smooth space curve, i.e. a curve in three-dimensional space. It has the property that the tangent line at any point makes a constant angle with a fixed line called the *axis*. Examples of helixes are coil springs and the handrails of spiral staircases. A "filled-in" helix – for example, a spiral ramp – is called a helicoid. Helices are important in biology, as the DNA molecule is formed as two intertwined helices, and many proteins have helical substructures, known as alpha helices. The word *helix* comes from the Greek word *λéð*, "twisted, curved".

Types

Helixes can be either right-handed or left-handed. With the line of sight along the helix's axis, if a clockwise screwing motion moves the helix away from the observer, then it is called a right-handed helix; if towards the observer, then it is a left-handed helix. Thus, a helix cannot be described as 'spinning clockwise or counter-clockwise.' Handedness (or chirality) is a property of the helix, not of the perspective: a right-handed helix cannot be turned or flipped to look like a left-handed one unless it is viewed in a mirror, and vice versa.

Most hardware screw threads are right-handed helices. The alpha helix in biology as well as the A and B forms of DNA are also right-handed helices. The Z form of DNA is left-handed.

The pitch of a helix is the width of one complete helix turn, measured parallel to the axis of the helix.

A double helix consists of two (typically congruent) helices with the same axis, differing by a translation along the axis.

A conic helix may be defined as a spiral on a conic surface, with the distance to the apex an exponential function of the angle indicating direction from the axis. An example is the Corkscrew roller coaster at Cedar Point amusement park.

A circular helix, (i.e. one with constant radius) has constant band curvature and constant torsion.

A curve is called a general helix or cylindrical helix if its tangent makes a constant angle with a fixed line in space. A curve is a general helix if and only if the ratio of curvature to torsion is constant.

A curve is called a slant helix if its principal normal makes a constant angle with a fixed line in space. It can be constructed by applying a transformation to the moving frame of a general helix.

Mathematical Description

In mathematics, a helix is a curve in 3-dimensional space. The following parametrisation in Cartesian coordinates defines a helix:

$$x(t) = \cos(t),$$
$$y(t) = \sin(t),$$
$$z(t) = t.$$

As the parametre t increases, the point $(x(t),y(t),z(t))$ traces a right-handed helix of pitch 2π and radius 1 about the z-axis, in a right-handed coordinate system.

In cylindrical coordinates (r, θ, h), the same helix is parametrised by:

$$r(t) = 1,$$
$$\theta(t) = t,$$
$$h(t) = t.$$

A circular helix of radius a and pitch $2\pi b$ is described by the following parametrisation:

$$x(t) = a\cos(t),$$
$$y(t) = a\sin(t),$$
$$z(t) = bt.$$

Another way of mathematically constructing a helix is to plot a complex valued exponential function (e^{xi}) taking imaginary arguments.

Except for rotations, translations, and changes of scale, all right-handed helices are equivalent to the helix defined above. The equivalent left-handed helix can be constructed in a number of ways, the simplest being to negate any one of the x, y or z components.

Arc length, curvature and torsion

The length of a circular helix of radius a and pitch $2\pi b$ expressed in rectangular coordinates as

$$t \mapsto (a\cos t, a\sin t, bt), t \in [0,T]$$

equals $T \cdot \sqrt{a^2 + b^2}$, its curvature is $\frac{|a|}{a^2 + b^2}$ and its torsion is $\frac{b}{a^2 + b^2}$.

Developable Surface

In mathematics, a developable surface (or torse: archaic) is a surface with zero Gaussian curvature. That is, it is a "surface" that can be flattened onto a plane without distortion (i.e. "stretching" or "compressing"). Conversely, it is a surface which can be made by transforming a plane (i.e. "folding", "bending", "rolling", "cutting" and/ or "gluing"). In three dimensions all developable surfaces are ruled surfaces. There are developable surfaces in R^4 which are not ruled.

Particulars

The developable surfaces which can be realized in three-dimensional space include:

- Cylinders and, more generally, the "generalized" cylinder; its cross-section may be any smooth curve
- Cones and, more generally, conical surfaces; away from the apex
- The oloid is one of very few geometrical objects that develops its entire surface when rolling down a flat plane.
- Planes (trivially); which may be viewed as a cylinder whose cross-section is a line
- Tangent developable surfaces; which are constructed by extending the tangent lines of a spatial curve.
- The torus has a metric under which it is developable, but such a torus does not embed into 3D-space. It can, however, be realized in four dimensions.

Formally, in mathematics, a developable surface is a surface with zero Gaussian curvature. One consequence of this is that all "developable" surfaces embedded in 3D-space are ruled surfaces (though hyperboloids are examples of ruled surfaces which are not developable). Because of this, many developable surfaces can be visualised as the surface formed by moving a straight line in space. For example, a cone is formed by keeping one end-point of a line fixed whilst moving the other end-point in a circle.

Application

Developable surfaces have several practical applications. Many cartographic projections involve projecting the Earth to a developable surface and then "unrolling" the surface into a region on the plane. Since they may be constructed by bending a flat sheet, they are also important in manufacturing objects from sheet metal, cardboard, and plywood (an industry which uses developed surfaces extensively is shipbuilding).

Non-developable Surface

Most smooth surfaces (and most surfaces in general) are not developable surfaces. Non-developable surfaces are variously referred to as having "double curvature", "doubly curved", "compound curvature", "non-zero Gaussian curvature", etc.

Some of the most often-used non-developable surfaces are:

- Spheres are not developable surfaces under any metric as they cannot be unrolled onto a plane.
- The helicoid is a ruled surface – but unlike the ruled surfaces mentioned above, it is not a developable surface.
- The hyperbolic paraboloid and the hyperboloid are slightly different doubly-ruled surfaces – but unlike the ruled surfaces mentioned above, neither one is a developable surface.

Applications of non-developable surfaces

Many gridshells and tensile structures and similar constructions gain strength by using (any) doubly curved form.

Bibliography

Ablameyko SV: *Recognition of Graphic Images*, Institute of Engineering Cybernetics, Minsk, 1996.

Agam G, Luo H and Dinstein I.: *Graphics Recognition — Methods and Applications*, Springer, Berlin, Springer, Berlin, 1996.

Baird HS: *Proc. of IAPR Workshop on Syntactic and Structural Pattern Recognition*, Murray Hill, NJ, 1990.

Basant Agrawal and C M Agrawal: *Engineering Drawing*, Tata McGraw Hill, New Delhi, 2008.

Charles R and Richard CW: *The Programmer's Apprentice*, ACM Press, Reading, MA, 1990.

Chung K and Yoon J.: *Proc. of International Conference on Document Analysis and Recognition*, Ulm, Germany, August, 1997.

Coad P and Yourdon E.: *Object-Oriented Analysis*, Yourdon Press, Prentice-Hall, Englewood Cliffs, NJ, 1990.

Das AK and Langrana NA: *Proc. of Int. Conf. on Document Analysis and Recognition*, Montreal, Canada, 1995.

De Marco T: *Structured Analysis and System Specification*, Yourdon Press, New York, 1978.

Dori D, Phillips I, and Haralick RM: *Applications of Object-Oriented Technology in Manufacturing*, Chapman Hall, Landon, 1994.

French, Thomas E.: *A Manual of Engineering Drawing for Students and Draftsmen,* McGraw-Hill, New York, USA, 1918.

Gao J, Tang L Liu W, and Tang Z: *International Conference on Document Analysis and Recognition*, Montreal, Canada, 1995.

Haralick RM and Shapiro L: *Computer and Robot Vision*, Addison Wesley, Reading, MA, 1992.

Harris JF, Kittler J, Llewellyn B, and Preston G: *Pattern Recognition: Theory and Applications*, D. Reidel Publisher, Norwell, Mass., 1982.

Iwama T, Horiuchi T, Toraichi K, Yamada H, and Yamamoto K: *IAPR Workshop on Machine Vision Applications*, Kawasaki, Japan, 1994.

Jacobson I, Christensson M, Jonsson P, and Overgaard GG: *Object-Oriented Software Engineering*, Addison-Wesley, Reading, MA, 1992.

Kadonaga T and Abe K.: *Graphics Recognition — Methods and Applications,* Springer, Berlin, 1996.

Kanungo T, Baird HS, and Haralick RM: *Fourth Annual Symposium on Document Analysis and Information Retrieval*, Las Vegas, Nevada, 1995.

Lee S, Lam L, Suen CY: T*he first International Conference on Document Analysis and Recognition*, Saint-Malo, France, 1991.

Madej D and Sokolowski A: *Int. Conf. on Document Analysis and Recognition,* Tsukuba, Japan, 1993.

Nalwa VS: *A Guided Tour of Computer Vision*, Addison-Wesley, New York, 1993.

O'Gorman L and Kasturi R: *Document Image Analysis*, IEEE Computer Society Press, Los Alamitos, CA, 1995.

Rumbaugh J, Blaha M, Premerlani W, Eddy F, and Lorensen W: *Object Oriented Modeling and Design*, Prentice Hall International Editions, Englewood Cliffs, NJ, 1991.

Sidheswar N, Kannaiah P and Sastry VVS: *Machine Drawing*, Tata McGraw-Hill Publishing Co. Ltd., New Dehli, India, 1980.

Sugihara K: *Machine Interpretation of of Line Drawings*, MIT Press, London, 1986.

Tombre K and Dori D: *Handbook of Character Recognition and Document Image Analysis*, World Scientific Publishing Company, 1997.

Waltz DA: *The Psychology of Computer Vision*, McGraw-Hill, New York, 1975.

Zhou Q and Haralick RM: *Experimental Protocol for Groundtruthing Line Drawings*, University of Washington, Seattle, WA, 1995.

Index

❑❑❑